FOUNDATIONAL MATHEMATICS

$$ax^2 + bx + c = 0$$

$$\sin^2 \theta + \cos^2 \theta = 1$$

$$\int_a^b f(x)dx = F(b) - F(a)$$

Ben McGahee, M.A.

Professional Mathematics Tutor

Former Adjunct Professor of Mathematics

REVIEW #1

THE REAL NUMBERS

The real numbers are comprised of a few important sets that the student has learned in mathematics.

Natural Numbers: The positive whole numbers used in the natural world such as 1, 2, 3,...

Integers: The Natural Numbers and their opposites, (including zero)..., -3, -2, -1, 0, 1, 2, 3,...

Rational Numbers: Any ratio of two integers so that the denominator is not zero such as 1/2, -5/4, repeating and terminating decimals like 0.333... and 0.7895 respectively.

Irrational Numbers: Numbers that are not rational such as $\sqrt{2}$, π, and e.

Recall that $\pi = 3.14159...$ and $e = 2.71828...$

PROPERTIES OF THE REAL NUMBERS

Here are some important properties of the real numbers that the student should already be familiar with.

For any real numbers x, y, and z

Commutative For + and *: $x + y = y + x$, $xy = yx$

Associative For + and *: $x + (y + z) = (x + y) + z$, $x(yz) = (xy)z$

Identity: $x + 0 = x$, where 0 is the additive identity, $1x = x$, where 1 is multiplicative identity

Also, note that $x + (-x) = 0$ where $-x$ is the additive inverse, and $x*1/x = 1$ for any non-zero x, where $1/x$ is the multiplicative inverse.

Distributive: $x(y + z) = xy + xz$

Additive Equality: If $x = y$, then $x + z = y + z$ for any real number z.

Multiplicative Equality: If $x = y$, then $xz = yz$ for any real number z.

It is important to note that division is the same operation as multiplication of the inverse. Division by zero is an undefined operation.

Additive Inequality: If $x < y$, then $x + z < y + z$ for any real number z.

Multiplicative Inequalities:

    A. If $x < y$ and $z > 0$, then $xz < yz$. This is true for $\leq$.
    B. If $x > y$ and $z > 0$, then $xz > yz$. This is true for $\geq$.
    C. If $x < y$ and $z < 0$, then $xz > yz$. This is true for $\leq$.
    D. If $x > y$ and $z < 0$, then $xz < yz$. This is true for $\geq$.

Interval Notation:

$x < a$ means $(-\infty, a)$
$x \leq a$ means $(-\infty, a]$
$x > a$ means (a, ∞)
$x \geq a$ means $[a, \infty)$
$a < x < b$ means (a, b) Also known as an open interval.
$a \leq x \leq b$ means $[a, b]$ Also known as a closed interval.
$a \leq x < b$ means $[a, b)$ Also known as a semi-open/closed interval.
$a < x \leq b$ means $(a, b]$. Also known as a semi-open/closed interval.
The set of real numbers $\mathbf{R} = (-\infty, \infty)$

Absolute Value Properties:

Absolute Value: $|x| = x$, if $x \geq 0$ and $|x| = -x$, if $x < 0$. Note that the definition of absolute value is the distance a number is from 0 or the origin and $|0| = 0$.

Triangle Inequality: $|x + y| \leq |x| + |y|$

Multiplication: $|xy| = |x||y|$

Division: $|x / y| = |x| / |y|$, where y is non-zero.

PROPERTIES OF EXPONENTS

The student should recall that an exponent n of any non-zero base a tells us that we should multiply a n number of times. That is $a^n = a*a*a\ldots(n\text{-times})$. Below are some important properties of exponents.

For any non-zero real base a and all real numbers m and n, we have the following:

Multiplication: $a^m a^n = a^{m+n}$

Division: $a^m / a^n = a^{m-n}$

Power to Power: $(a^m)^n = a^{mn}$

Zero Exponent: $a^0 = 1$

Negative Exponent: $a^{-n} = 1 / a^n$

Two Bases to Power: $(ab)^n = a^n b^n$

Division to Power: $(a / b)^n = a^n / b^n$

Division with Negative Exponent: $(a / b)^{-n} = (b / a)^n$

PROPERTIES OF RADICALS

Here are some familiar properties of radicals that the student should know such as how to convert a radical to a rational exponent and vice versa. Also, the addition, subtraction, multiplication, and division rules for radicals are pretty straightforward.

For any bases x and y, and $n \geq 2$ we have the following:

Radical to Rational Exponent: $\sqrt[n]{x} = x^{1/n}$

General Property: $\sqrt[n]{x^m} = x^{m/n} = \left(\sqrt[n]{x}\right)^m$, where m is any integer.

Addition: $a\sqrt[n]{x} + b\sqrt[n]{x} = (a+b)\sqrt[n]{x}$, for any real constants a and b.

Subtraction: $c\sqrt[n]{y} + d\sqrt[n]{y} = (c+d)\sqrt[n]{y}$, for any real constants c and d.

Multiplication: $\sqrt[n]{x}\sqrt[n]{y} = \sqrt[n]{xy}$

Division: $\sqrt[n]{\dfrac{x}{y}} = \dfrac{\sqrt[n]{x}}{\sqrt[n]{y}}$, where y is non-zero.

PROPERTIES OF FACTORING

The student should be aware how to factor polynomials such as binomials, trinomials, sum and difference of cubes, etc. from elementary algebra. Recall that factoring is writing an algebraic expression as a product of algebraic expressions. Also, a polynomial that cannot be factored is considered "prime."

Greatest Common Factor: $xy + xz = x(y + z)$, where x is the greatest common factor. This is like doing distributive property in reverse.

Factor by Grouping: $ax + bx + ay + by = x(a + b) + y(a + b) = (a + b)(x + y)$.

Factor Trinomial Case 1: $x^2 + bx + c$

-Find factors of constant term c that add up to the constant b.

Factor Trinomial Case 2: $ax^2 + bx + c$ This approach is called the "Grouping Method"

-Multiply constants a and c to get ac.

-Find factors of ac that add up to b.

-Use Factor by Grouping Property

Sum of Squares: $x^2 + y^2$ is prime with real numbers.

Difference of Two Squares: $x^2 - y^2 = (x + y)(x - y)$

Sum of Cubes: $x^3 + y^3 = (x + y)(x^2 - xy + y^2)$

Difference of Cubes: $x^3 - y^3 = (x - y)(x^2 + xy + y^2)$

LINEAR EQUATIONS

A linear equation can be written in the form $ax + b = c$, where a, b, and c are real numbers and a is non-zero. The variable x is the solution to a linear equation.

In a graphical context, the student should be familiar with the following forms of a line.

Slope-Intercept Form: $y = mx + b$, where m is the slope and b is the y-intercept.

Point-Slope Form: Let (x_1, y_1) be a point on the line. Then $y - y_1 = m(x - x_1)$.

Standard Form: $ax + by = c$, where a, b, and c are real numbers and either a or b (not both) can be zero.

Important Definitions:

Slope: If (x_1, y_1) and (x_2, y_2) are points on a line, then the slope $m = \dfrac{y_2 - y_1}{x_2 - x_1}$.

Y-Intercept: The point $(0, y)$ where the graph crosses the y-axis.

X-Intercept: The point $(x, 0)$ where the graph crosses the x-axis.

Parallel Lines: Two lines are parallel if they have the same slope.

Perpendicular Lines: Two lines are perpendicular if the product of their slopes is -1.

Lines that are neither parallel nor perpendicular are known as "skew lines."

HOW TO GRAPH A LINEAR EQUATION IN SLOPE-INTERCEPT FORM:

1. Identify the slope and y intercept in the linear equation $y = mx + b$.

2. Graph the point $(0, b)$ located on the y-axis.

3. Use the slope m = rise / run to go up or down y-units (rise) and left or right x-units (run) to plot another point.

4. Repeat step 3.

5. Create a line by connecting all 3 points.

SYSTEMS OF LINEAR EQUATIONS

The student should be able to identify a system of linear equations and know various ways of solving them.

A system of linear equations is two or more equations involving two or more variables.

A system involving lines in the plane contain only two variables x and y.

A system involving planes contain three variables x, y, and z.

TYPES OF SYSTEMS:

A. Consistent: System has at least one solution.

B. Inconsistent: System has no solution.

C. Dependent: System whose equations are multiples of one another or "depend" on one another. E.g. Coinciding lines and planes

D. Independent: System whose equations are not multiples of one another.

E.g. Perpendicular Lines

METHODS OF SOLVING SYSTEMS:

1. Graphing

2. Substitution

3. Elimination

QUADRATIC EQUATIONS

A quadratic equation in standard form is a second degree polynomial equation that can be written as $ax^2 + bx + c = 0$, where a, b, and c are real numbers and a is non-zero. There are three important methods the student knows in solving quadratic equations.

1. Factoring

2. Completing the Square

3. Quadratic Formula (This results from completing the square).

Note: It is important to know that not all quadratic equations can be factored, so the only methods that we can apply are 2. and 3.

STEPS FOR COMPLETING THE SQUARE:

1. Write quadratic equation in standard form.

2. Bring constant term c to other side of equation.

3. Divide by a to both sides of equation.

4. Take half of the middle term to get $b/2a$.

5. Square $b/2a$, which is $b^2/4a^2$.

6. Add result of 5. to both sides.

7. Factor the perfect square trinomial on the left side and combine terms on the right side.

8. Square root both sides of the equation.

9. Solve for x and simplify.

THE QUADRATIC FORMULA: WEAPON OF MATH INSTRUCTION

$$x = \frac{-b \pm \sqrt{b^2 - 4ac}}{2a}$$

This formula solves any quadratic equation $ax^2 + bx + c = 0$.

An important feature of the quadratic formula is the expression located underneath the square root, which is called the "discriminant."

The discriminant $D = b^2 - 4ac$. Let us consider 3 different cases for the discriminant.

1. If $D > 0$, then we have two distinct real roots.

2. If $D = 0$, then we have one real root.

3. If $D < 0$, then we have two complex roots.

It follows that case 3 is the only case where we have no real roots.

THE COMPLEX NUMBERS

The student should recall that it is impossible to take the square root (or any even root) of a negative number. Mathematicians solved this problem by introducing the imaginary number. Here are the four imaginary numbers we should all be familiar with.

$$i = \sqrt{-1}$$
$$i^2 = -1$$
$$i^3 = -i$$
$$i^4 = 1$$

A complex number is written in the form $z = a + bi$, where a is the real part of z or Re(z) and

b is the imaginary part of z or Im(z). Like the real numbers, we can add, subtract, multiply, and

divide complex numbers. A complex number also has a special feature called the conjugate z^*,

where $z^* = a - bi$.

Here are the methods we use for the four operations.

Let z_1 and z_2 be complex numbers.

Addition: Add the real parts together and the imaginary parts together of z_1 and z_2.

Subtraction: Take the opposite of z_2 and add the real and imaginary parts to z_1.

Multiplication: Use the FOIL (first, outside, inside last) method to multiply the complex numbers and combine the real and imaginary parts of z_1 and z_2.

Division: The way to simplify z_1 / z_2 is to multiply the top and bottom by the conjugate of z_2.

It can be shown that the product of a complex number and its conjugate results in a real number.

Another important aspect of the complex number is its distance from the origin in the plane. This distance is known as the modulus of z or $|z|$ (read as "mod z"). Below is the definition.

$$|z| = \sqrt{a^2 + b^2} \ .$$

We will return to the topic of complex numbers after we establish some ideas in trigonometry and calculus.

COMPUTATIONAL EXERCISES #1

1. Determine which set(s) each of the follow numbers belong.

    a) -3

    b) 5.7

    c) $\pi + 1$

    d) 4/9 – 7/18

    e) $2 + 6i$

2. Use the laws of exponents to simplify each expression. Write each answer in terms of positive exponents.

    a) $2x^4 y^3 x^{-7}$

    b) $(5a^2 b)^{-3}$

    c) $18m^8 n^{-12} p^6 / 10m^0 n^9 p^{-7}$

    d) $(ab^2 / a^{-1} b)^4 a^{-8} b^{-6}$

3. Simplify each of the following radical expressions.

   a) $\sqrt{60x^2 y^4}$

   b) $\sqrt{\dfrac{81}{25}}$

   c) $3\sqrt{2x} - 4\sqrt{18x} + 20\sqrt{72x^3}$

   d) $729^{4/3} + 256^{5/2}$

4. Factor each of the following polynomials.

   a) $2x^2 + 6x$

   b) $4y^3 + 8y + 16y^3 + 32y$

   c) $t^2 - 10t - 96$

   d) $7a^4 - 343b^4$

   e) $m^6 + n^6$

5. Solve each of the following problems with linear equations.

   a) $5(2x - 4) + 6x + 8 = -9x - 7$

   b) $2y + 3 = 4(y/2 - 1/8) + 1/2$

   c) $1/6(12 - 24z) + 4z = 2$

   d) Find an equation of a line passing through (-3, 7) and (8, 10).

   State the slope along with the x and y intercepts. Graph the line.

   e) Determine whether the lines are parallel, perpendicular, or neither.

   $3x + 6y = 12, y = 2x - 4$

   f) Solve the system of equations by elimination. $10a + 5b = -20, 2b = 5a + 40$

6. Solve each of the following problems with quadratic equations.

   a) $x^2 - 7x + 12 = 0$

   b) Solve by completing the square. $t^2 + 4t - 8 = 0$

   c) Solve by the quadratic formula. $r(r - 2) + 2r^2 = 3r + 6$

   d) Use the discriminant to determine how many real or complex roots exist in the equation $8w^2 = 5w - 1$.

7. Simplify each of the following problems with complex numbers. Write in $a + bi$ form.

   a) $-2i^6 + 4i + 3 - 7i$

   b) $(5 + 10i)(4 - 9i) + 20 - 30i$

   c) $(1 + i) / (-1 - 2i)$

   d) Find $|z|$ for each complex number in a), b), and c).

CONCEPTUAL EXERCISES #1

1. True or False "Every integer is also a rational number."

2. True or False "It is possible to take the even root of any real number."

3. True or False "A system of linear equations that contains parallel lines is a consistent system."

4. True or False "A positive discriminant indicates that a quadratic equation has two real roots."

5. True or False "The quotient of a complex number and its conjugate generates a real number."

6. If two points (a, b) and (b, a) form a line, where a and b are non-zero real numbers, then what is the sign of the slope?

7. If $c = pq$ and $b = p + q$, then what is the factorization of $x^2 + bx + c$ in terms of x, p, and q?

8. Derive the quadratic formula from the equation $ax^2 + bx + c = 0$.

9. If the parabola $y = x^2 + bx + c$ has only one x-intercept and $c = b/4$, find two quadratic equations that fit this criteria.

10. Let z_1 and z_2 be two, distinct complex numbers.

   If $z_1' = z_1 / |z_1|^2$ and $z_2' = z_2 / |z_2|^2$, show that $z_1 / z_2* + z_2 / z_1*$ can be written in the form $Rz_1'z_2'$, where R is a positive real number.

REVIEW #2

RELATIONS AND FUNCTIONS

In algebra, we know that a relation R is a set of ordered pairs on a given set S. There are 3 important relations: 1. Reflexive 2. Symmetric 3. Transitive

Here are their definitions below.

Let R be a relation on a given set S.

1. Reflexive: For any x in S, (x, x) is in R.

2. Symmetric: For any x, y in S, if (x, y) is in R then (y, x) is in R.

3. Transitive: For any x, y, z in S, if (x, y) is in R and (y, z) is in R then (x, z) is in R.

A relation that is reflexive, symmetric, and transitive is called an "equivalence relation."

We know that some relations are special and possess certain rules or operations that one should follow. These relations are known as "functions."

A function is a special correspondence between two sets: A. Domain and B. Range

In order for a relation to be a function, every element in the domain must correspond to exactly one element in the range.

If we consider a function from a computer perspective, we notice that the domain contains all of the input values and the range contains all the output values that result from applying the rule of the function.

That is, x is an element in the domain, y is an element in the range and f is the function.

The function notation that all students should be familiar with is $y = f(x)$. This is read as "y equals f of x." In other words, f is a function of x. The independent variable is x, and the dependent variable is y or $f(x)$.

How can one determine if a graph represents a function or not? There is an important test we can use to find this out.

VERTICAL LINE TEST: Draw a vertical line through any given point on the graph. If the vertical line touches the graph exactly once, then the graph represents a function.

OPERATIONS AND PROPERTIES OF FUNCTIONS

The student should know the following operations of functions.

Let f and g be two functions.

ADDITION: $(f + g)(x) = f(x) + g(x)$

SUBTRACTION: $(f - g)(x) = f(x) - g(x)$

MULTIPLICATION: $(fg)(x) = f(x)g(x)$

DIVISION: $(f / g)(x) = f(x) / g(x)$, where $g(x)$ is non-zero

Functions take on important properties such as symmetry, shifting, stretching, and shrinking.

SYMMETRY

EVEN FUNCTION: If $f(-x) = f(x)$, then f is even and is symmetric about the y-axis.

ODD FUNCTION: If $f(-x) = -f(x)$, then f is odd and is symmetric about the origin.

SHIFTING

HORIZONTAL SHIFT: $f(x - h)$ means to shift $f(x)$ h units horizontally.

1. If $h > 0$, then f is shifted h units to the right.

2. If $h < 0$, then f is shifted h units to the left.

VERTICAL SHIFT: $f(x) + k$ means to shift $f(x)$ k units vertically.

1. If $k > 0$, then f is shifted k units upward.

2. If $k < 0$, then f is shifted k units downward.

HORIZONTAL STRETCHING: If $0 < c < 1$, then $f(cx)$ means to stretch $f(x)$ horizontally by a factor of c.

VERTICAL STRETCHING: If $c > 1$, then $cf(x)$ means to stretch $f(x)$ vertically by a factor of c.

HORIZONTAL SHRINKING: If $c > 1$, then $f(cx)$ means to shrink $f(x)$ horizontally by a factor of c.

VERTICAL SHRINKING: If $0 < c < 1$, then $cf(x)$ means to shrink $f(x)$ vertically by a factor of c.

LINEAR FUNCTIONS

A function written in the form $f(x) = mx + b$, where m is the slope and b is the y-intercept is considered to be a linear function. The graph of a linear function represents a straight line.

DOMAIN: The domain for any linear function (unless piecewise restricted) is all real numbers.

RANGE: The range depends on a couple of cases.

Case 1: If m is non-zero, then the range is all real numbers.

Case 2: If m is zero, then the range is $\{b\}$ or the y-intercept.

NON-LINEAR FUNCTIONS

A function that is not linear or cannot be written in $f(x) = mx + b$ is called a non-linear function. Such functions can be polynomial functions such as quadratic, cubic, quartic, quantic, etc. as well as absolute value, root, rational, exponential, trigonometric, and so forth. The graphs of these functions represent curves instead of straight lines. Below are some categories of non-linear functions.

POLYNOMIAL FUNCTIONS

A polynomial function can be written in the form $p(x) = a_n x^n + a_{n-1} x^{n-1} + \ldots + a_1 x + a_0$, where the coefficients $a_0, a_1, \ldots, a_n$ are real numbers and n is the degree of the polynomial. Here are some types of polynomial functions with various degrees.

Zero Degree: Constant Function

First Degree: Linear Function

Second Degree: Quadratic Function

Third Degree: Cubic Function

Fourth Degree: Quadratic Function

Fifth Degree: Quintic Function

QUADRATIC FUNCTIONS

A quadratic function can be written in a couple of forms: 1. Standard 2. Vertex

The graph of any quadratic equation is called a parabola.

STANDARD: $f(x) = ax^2 + bx + c$, where a is non-zero.

VERTEX: $f(x) = a(x - h)^2 + k$, where (h, k) is the vertex.

Recall that the vertex is the point when the parabola opens upward or downward, depending on the value of a.

Parabola opens upward if $a > 0$.

Parabola opens downward if $a < 0$.

The axis of symmetry is the vertical axis that cuts the parabola into two equal parts.

The equation for the axis of symmetry is $x = h$.

A quadratic equation can have a minimum or a maximum value depending on the value of a.

If $a > 0$, then f has a minimum value of k.

If $a < 0$, then f has a maximum value of k.

VERTEX FORMULA: If the quadratic equation is written in standard form, the x and y coordinates of the vertex can be found by the following.

$x = -b / 2a$

$y = f(-b / 2a)$ Plug the x coordinate of the vertex into the quadratic function to find the y coordinate of the vertex.

ABSOLUTE VALUE FUNCTIONS

An absolute value function can be written as $f(x) = |x|$. We can think of the absolute value function as a piecewise function since it can be broken up into two different functions for different parts of their domain.

1. $f(x) = x$ if $x \geq 0$

2. $f(x) = -x$ if $x < 0$

DOMAIN: All real numbers

RANGE: All non-negative real numbers

ROOT FUNCTIONS

A root function can be written as $f(x) = x^{1/n}$ in exponential form or equivalently in radical form as $f(x) = \sqrt[n]{x}$, where n is a natural number such that $n \geq 2$. The domain and range of a root functions depends on whether n is even or odd.

DOMAIN:

If n is even, then the domain is all non-negative real numbers.

If n is odd, then the domain is all real numbers.

RANGE:

If n is even, then the range is all non-negative real numbers.

If n is odd, then the range is all real numbers.

RATIONAL FUNCTIONS

A rational function can be written as $r(x) = f(x) / g(x)$, where $f(x)$ and $g(x)$ are polynomial functions and $g(x)$ is non-zero. There are two important pieces to rational functions:

1. Vertical Asymptote

2. Horizontal Asymptote

HOW TO FIND THE VERTICAL ASYMPTOTE

1. Set $g(x) = 0$.

2. Find the value(s) of x that satisfy $g(x) = 0$.

If a is the value such that $g(a) = 0$, then a vertical dashed line will be drawn at $x = a$ to indicate that $r(x)$ is undefined at a.

HOW TO FIND THE HORIZONTAL ASYMPTOTE

1. Check the degree of $f(x)$ and $g(x)$ and consider 3 cases

A. If the degrees of $f(x)$ and $g(x)$ are equal, then the graph of $r(x)$ will approach a certain number r on the y-axis as x approaches ∞ or $-\infty$. The number r is the ratio of leading coefficients of $f(x)$ and $g(x)$. The equation $y = r$ is the horizontal asymptote.

B. If the degree of $f(x)$ is one less than the degree of $g(x)$, then the horizontal asymptote equation is $y = 0$.

C. If the degree of $f(x)$ is one more than the degree of $g(x)$, then the asymptote is not horizontal, but slant/oblique. The equation for the slant/oblique asymptote is $y = ax + b$, where a is the slope and b is the y-intercept.

The domain and range of a rational function depend mostly on the vertical and horizontal asymptotes. In general, we have the following.

DOMAIN: All real numbers except value(s) of vertical asymptote(s).

RANGE: All real numbers except value of horizontal asymptote.

We will discuss important aspects of exponential and trigonometric functions in the near future.

COMPOSITION OF FUNCTIONS

Two or more functions can be added, subtracted, multiplied, and divided, but they can also be composed. This means that one function can be inputted into another function to generate a new function as a result. Below is the standard notation for composition of functions.

$(f \circ g)(x) = f(g(x))$ This is read as "f of g of x." We input the function $g(x)$ into $f(x)$ and receive a new function $f(g(x))$.

DOMAIN: The range of $g(x)$

RANGE: The set of all output values of $f(g(x))$

$(g \circ f)(x) = g(f(x))$ This is read as "g of f of x." We input the function $f(x)$ into $g(x)$ and receive a new function $g(f(x))$.

DOMAIN: The range of $f(x)$

RANGE: The set of all output values of $g(f(x))$

In general, we should note that composition of two functions is not commutative. That is $f \circ g \neq g \circ f$ for all functions f and g. We will investigate the case when the compositions are equal.

However, the composition of three functions is associative.

That is, $f \circ (g \circ h) = (f \circ g) \circ h$ for all functions f, g, and h.

ONE-TO-ONE AND INVERSE FUNCTIONS

We stated earlier that a function is when every x value in the domain corresponds to exactly one y value in the range. A one-to-one function is a special function that satisfies the following definition.

Definition of One-To-One: If $f(x) = f(y)$ then $x = y$.

This is equivalent to saying: If $x \neq y$, then $f(x) \neq f(y)$.

In other words, a one-to-one function has every y value in the range belonging to exactly one x value in the domain.

Like the vertical line test we applied to determine whether a graph is a function, we can apply a new test to determine whether the graph of a function is one-to-one. This is known as the horizontal line test.

HORIZONTAL LINE TEST: If a horizontal line can be drawn at any point on the graph so that it touches the graph only once, then the function is one-to-one.

A one-to-one function also contains an important component called the inverse function, which is denoted as $f^{-1}(x)$. The domain of the inverse function is the range of the original function. The range of the inverse function is the domain of the original function. In other words, the domain and range of the original function swap places for the inverse function.

When you compose a function and its inverse (in either order) you will get the same result. This means that $f(f^{-1}(x)) = f^{-1}(f(x)) = x$, where x is called the identity function. On a geometric level, we observe that the graph of $f(x)$ is reflected over the line $y = x$ to create the graph of $f^1(x)$ and vice versa.

HOW TO FIND THE INVERSE FUNCTION

Here are a few steps on finding the inverse function, provided that it is one-to-one and passes the horizontal line test.

1. Replace $f(x)$ with y.

2. Switch the variables x and y.

3. Solve for y in the new equation.

4. Replace y with $f^1(x)$. This is your inverse function.

EXPONENTIAL AND LOGARITHMIC FUNCTIONS

An exponential function involves a positive base, which is raised to a variable exponent. Let us review the different types of exponential functions.

A. $f(x) = a^x$ We consider a couple of cases.

1. If $a > 1$, then f is increasing exponentially. This is known as exponential growth.

2. If $0 < a < 1$, then f is decreasing exponentially. This is known as exponential decay.

DOMAIN: All real numbers

RANGE: All positive real numbers

The exponential function has no x-intercept, but a y-intercept $(0, 1)$.

As the value of x approaches negative infinity for case 1, we see that the value of y approaches 0. Therefore, we have a horizontal asymptote $y = 0$. Similarly, the value of y approaches 0 as x approaches positive infinity. We still have a horizontal asymptote $y = 0$.

B. $f(x) = e^x$ Recall that e is an irrational number which is about 2.71828... The number e in the context of exponential functions is known as the natural base. Like the previous exponential function, e^x has the same domain, range, y-intercept, and horizontal asymptote. This function has numerous applications in business, calculus, engineering, physics, etc.

If we apply the horizontal line test to any exponential function, we notice that the line touches the graph only once at each point. We can deduce that the function is one-to-one and has an inverse. But the inverse function of the exponential function introduces a new notation known as the logarithm.

We can apply the 4 easy steps to find the inverse of the exponential function $f(x) = a^x$.

1. $y = a^x$

2. $x = a^y$

3. Solve for y. But how? The solution is to introduce the logarithm.

$y = log_a(x)$ This is read as "log base a of x."

4. $f^{-1}(x) = log_a(x)$

So, the inverse function of the exponential function is a logarithmic function.

Here are some important facts to know about logarithmic functions.

-Base a is a natural number greater than 1.

-If $a = 10$, then the notation is written as $f(x) = log(x)$ with an implied base 10. This is known as the standard logarithm with a standard base 10.

-DOMAIN: All positive real numbers

-RANGE: All real numbers

-x-intercept of $(1, 0)$

-No y-intercept

-vertical asymptote at $x = 0$

-Many people abbreviate logarithm as "log."

THE NATURAL LOG FUNCTION

Recall that $f(x) = e^x$ is an exponential function that has a natural base e. Since e is the natural base, the inverse function would be a natural log function or $f^{-1}(x) = ln(x)$. The natural log function $ln(x)$ can also be written as $log_e(x)$. The domain, range, and x-intercept are the same as the original log function. Like the exponential function, the natural log function arises in many important applications such as calculus, engineering, physics, etc.

LOG PROPERTIES

We review the following log properties that are very helpful in simplifying log expressions and solving log equations.

1. Product-Sum Rule: $log_a(xy) = log_a(x) + log_a(y)$

2. Quotient-Difference Rule: $log_a(x/y) = log_a(x) - log_a(y)$

3. Power Rule: $log_a(x^p) = p log_a(x)$, where p is any real number

4. Base to Log Power: $a^{log_a(x)} = x$, where x is a positive real number

5. Log of a Base to Power: $log_a(a^x) = x$, where x is any real number

6. e to Natural Log Power: $e^{ln(x)} = x$, where x is any positive real number

7. Natural Log to e Power: $ln(e^x) = x$, where x is any real number

CHANGE OF BASE FORMULA

Sometimes, we may want to change the base of a logarithmic expression to make a simpler calculation. For example, we may want to change the log of base 3 to the log of base 10 since base 10 is the standard logarithm used on scientific and graphing calculators. Below is the change of base formula.

Change of Base: Let a be the original base and let b be the new base.

$log_a(x) = log_b(x) / log_b(a)$

All we are doing is taking the quotient of two separate log expressions of x and a with respect to the new base b.

We have witnessed several types of relations and functions (both linear and non-linear) that possess various domains, ranges, and certain properties. Functions can undergo transformations such as shifting, stretching, and shrinking in both the horizontal and vertical directions as well as having horizontal and vertical asymptotes. In calculus, we will have a better understanding of why functions behave certain ways and be able to decipher other important features on a more profound level.

COMPUTATIONAL EXERCISES #2

1. Determine whether the following relations represent a function. If the relation is a function, find the domain and range.

   a) $R = \{(1, 2), (3, 4), (-2, 1), (-5, 3), (6, 8)\}$

   b) $T = \{(a, \$), (b, !), (c, *), (b, @), (d, \&), (e, \#)\}$

   c) A line with a slope of -2 and a y-intercept of 1.

   d) A circle with a center at the origin and a radius of 5 units.

2. Classify each function as linear or non-linear and find the domain, range, x and y intercepts, vertex axis of symmetry, horizontal and/or vertical asymptotes (if any). Write the domain and range in interval notation. Also, graph each function.

   a) $f(x) = x^2 + 4x + 2$

   b) $f(x) = 5x - 3$

   c) $f(x) = |2x - 1|$

   d) $f(x) = -(1 + 2x)^{1/2}$

   e) $f(x) = \dfrac{x - 6}{x^2 - 9}$

   f) $f(x) = 2^{x-3} + 8$

   g) $f(x) = \ln(x + 7) - 4$

3. Let $f(x) = x^2 - 2x$, $g(x) = 3x^2 + x - 1$, and $h(x) = x^{1/2}$. Perform each of the following operations.

   a) $(f + g)(x)$

   b) $(g - f)(x)$

   c) $(fh)(x)$

   d) $(f / g)(2)$

e) $(g \circ h)(9)$

4. Find the inverse function for each of the following functions. State the domain and range.

   a) $f(x) = -6x + 12$

   b) $f(x) = x^3 - 8$

   c) $f(x) = 2e^{4x}$

   d) $f(x) = log_5(x + 1) - 3$

   e) $f(x) = \dfrac{x + 2}{2 - x}$

5. Simplify each of the logarithmic expressions.

   a) $log_2 8$

   b) $log_3 81^5$

   c) $log(60) - log(15) + log(25)$

   d) $log_a(xy) - log_a(y/z) + log_a(z) - log_a(x)$

6. Solve each of the following equations. Give exact answers.

   a) $5e^{3x} - 14 = 31$

   b) $4^{x-2} + 6 = 22$

   c) $ln(x - 2) + ln(x) = ln8$

   d) $log_2(x / (x + 1)) = -1$

CONCEPTUAL EXERCISES #2

1. True or False: Every relation is a function.

2. True or False: A function that passes the horizontal line is one-to-one and has an inverse.

3. True or False: A relation that is reflexive, symmetric, and transitive is an equivalence relation.

4. True or False: For any one-to-one function $f(x)$, $f^1(x) = 1 / f(x)$.

5. True or False: The standard base for $log(x)$ is base 10 and the natural base for $ln(x)$ is base e.

6. True or False: For any functions f and g, $f \circ g = g \circ f$.

7. True or False: The composition of functions f, g, and h is associative.

8. True or False: The logarithm of a product is the sum of the individual logarithms.

9. True or False: The logarithm of a quotient is the quotient of the individual logarithms.

10. True or False: $log_3 10$ can be written as $(log 3)^{-1}$.

11. Let Z be the set of integers and R be a relation on Z. (x, y) is in R if and only if $x \leq y$. Determine whether R is reflexive, symmetric, or transitive. Is R an equivalence relation?

12. Let Z be the set of integers and R be a relation on Z. (a, b) is in R if and only if

    $a \equiv b$ mod n. (Note that this means that n divides $a - b$ or $a - b = kn$ for some integer k).

    Prove that R is an equivalence relation.

13. For $f(x) = x^2 + bx + c$, if r_1 and r_2 are distinct, real roots, find the values of b and c in terms of r_1 and r_2.

14. Given that $f(x) = ax^2 + bx + c$, find the coordinates of the vertex (x, y) in terms of a, b, and c. Under what condition guarantees f a minimum or maximum value?

15. If $f(x) = e^x$, show that a) $f(-x) = 1 / f(x)$, b) $f(x + y) = f(x)f(y)$, c) $f(x - y) = f(x)/f(y)$, d) $f^1(xy) = f^1(x) + f^1(y)$, e) $f^1(x/y) = f^1(x) - f^1(y)$

16. If a and c are non-zero real numbers, show that $f(x) = ax^3 + c$ is one-to-one and find it inverse. Deduce that the y-intercept of f equals the x-intercept of f^1.

17. If f and g are one-to-one functions, prove that $f \circ g$ is also one-to-one.

18. A function is "onto" if and only if for any y in set B, there exists an x in set A such that $y = f(x)$. A function that is both one-to-one and onto is "bijective." Determine if the following function F is bijective.

$A = \{1, 2, 3, 4, 5\}$

$B = \{12, 23, 34, 45, 51\}$

$F = \{(1, 23), (3, 51), (4, 45), (2, 12), (5, 34)\}$

19. Let $f(x) = \dfrac{ax+b}{bx-a}$, where a and b are non-zero real numbers. If the horizontal asymptote is $y = ab$, what values of a and b would satisfy this condition? What would be the possible functions for each of these values?

20. Prove the following log properties.

 a) $log_b(xy) = log_b(x) + log_b(y)$

 b) $log_b(x/y) = log_b(x) - log_b(y)$

 c) $log_b(x^p) = plog_b(x)$

21. Show that $lne = 1$ and $e^{ln1} = 1$

22. The amount A of a substance that decays over a period of time is determine by the exponential formula $A = Ie^{rt}$, where I is the initial amount, r is the rate of decay, and t is the time. Show that the time it takes for the substance to decay to ½ of its original amount is $-ln2/r$. This is known as the half-life of a substance.

REVIEW #3

COORDINATE GEOMETRY

In this review, we are shifting gears to a more visual side of mathematics where the foundations of geometry are established. We will not investigate all topics of geometry or all the theorems one recalls in a high school or college setting. But we will touch base on some of the main ideas that will be useful for advanced topics in trigonometry and calculus.

THE XY-PLANE: Recall that the xy-plane is formed by two axes.

1. X-AXIS: The horizontal axis

2. Y-AXIS: The vertical axis

The point where the x-axis and y-axis meet is called the origin with coordinates (0, 0). The xy-plane is divided into 4 quadrants: I, II, III, and IV where the top right-hand quadrant is the first quadrant and the remaining quadrants go counterclockwise.

Any location in the plane is called a point and can be denoted as the ordered pair (x, y), where x is the number of units moved horizontally from the origin, and y is the number of units moved vertically from the origin. See the figure on the next page.

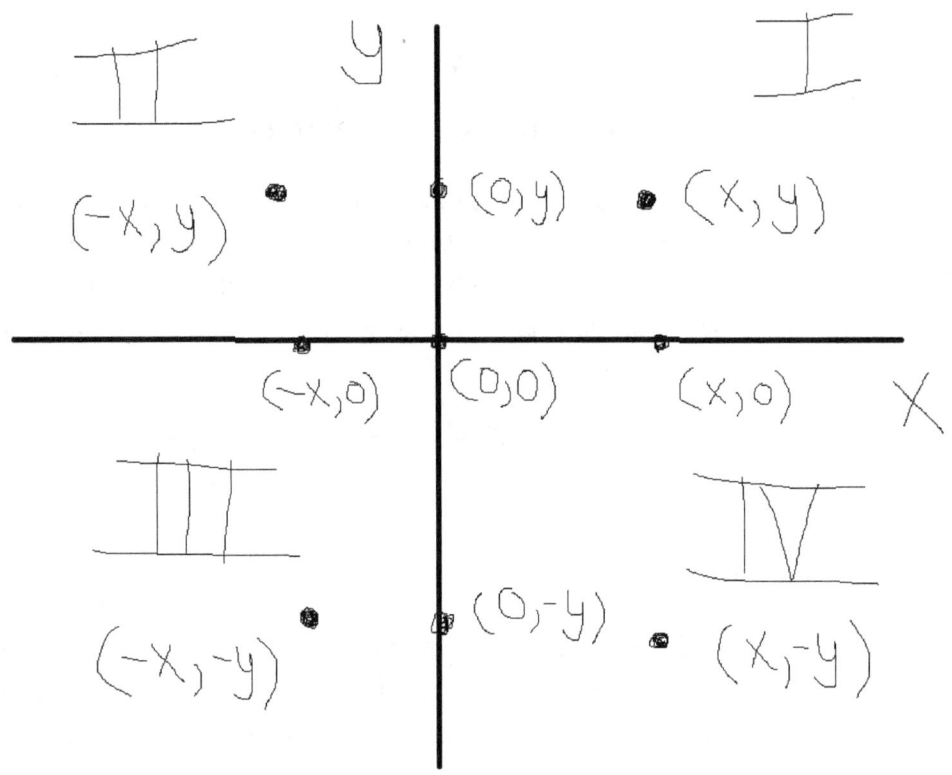

Recall the quadrants where each of the coordinates resides based on their given sign values.

1. If $x > 0$, $y > 0$, then (x, y) belongs in the 1st quadrant.

2. If $x < 0$, $y > 0$, then (x, y) belongs in the 2nd quadrant.

3. If $x < 0$, $y < 0$, then (x, y) belongs in the 3rd quadrant.

4. If $x > 0$, $y < 0$, then (x, y) belongs in the 4th quadrant.

5. Any point $(x, 0)$ belongs on the x-axis.

6. Any point $(0, y)$ belongs on the y-axis.

DISTANCE AND MIDPOINT FORMULAS

There are two important formulas in geometry that are very important if we want to find the distance between two points or find the middle point between two points. They are known as the 1. Distance Formula and 2. Midpoint Formula.

DISTANCE FORMULA: Let a first point P_1 have coordinates (x_1, y_1) and a second point P_2 have coordinates (x_2, y_2). Then the distance d between the two points P_1 and P_2 can be expressed by the following equation.

$$d = \sqrt{(x_2 - x_1)^2 + (y_2 - y_1)^2}$$

Note that distance is a non-negative number, which means that we can only have zero or positive distance. The reason this is true is due to the fact that the sum of two squares is either positive or zero. Also, common sense tells us that we travel zero distance if we stay at home (the origin) and don't move anywhere or we travel positive distance if we leave home to travel somewhere else. This equation is actually derived from the Pythagorean Theorem, which we will cover very soon.

MIDPOINT FORMULA: Let a first point P_1 have coordinates (x_1, y_1) and a second point P_2 have coordinates (x_2, y_2). Then the midpoint m between the two points P_1 and P_2 can be expressed by the following equation.

$$m = \left(\frac{x_1 + x_2}{2}, \frac{y_1 + y_2}{2} \right)$$

In order to find the x and y coordinates of the midpoint, we need to take the average of the x and y coordinates from both points respectively. It makes sense because taking the average of two numbers is finding the middle number, which is what we need to find the midpoint.

POLYGONS

Recall that a polygon is a closed geometric figure formed by line segments. Here are some examples of polygons.

TRIANGLE: A polygon with 3 sides

QUADRILATERAL: A polygon with 4 sides

PENTAGON: A polygon with 5 sides

HEXAGON: A polygon with 6 sides

OCTAGON: A polygon with 8 sides

N-GON: A polygon with N sides, where N $\geq$ 9.

One thing we can calculate of any polygon is the sum of the interior angles. This helps us to find the value of each individual angle formed by two adjacent sides.

SUM OF INTERIOR ANGLES

If n is the number of sides then the sum of the interior angles within the polygon is $S = (n - 2)180$.

To find the individual angle A, we use the formula $A = S/n$.

PYTHAGOREAN THEOREM

One of the most beautiful and famous theorems in geometry and in all of mathematics is the Pythagorean Theorem. The Pythagorean Theorem is named in honor of the Greek mathematician Pythagoras who made great contributions to the field of geometry. This theorem is used when we are dealing with right triangles, which are triangles that have a 90 degree (right) angle. A right triangle has two sides or legs and a hypotenuse, which is the longest side of the triangle opposite of the right angle. Below is the statement of the Pythagorean Theorem.

PYTHAGOREAN THEOREM: If a and b are legs of a right triangle and c is the hypotenuse, then $a^2 + b^2 = c^2$.

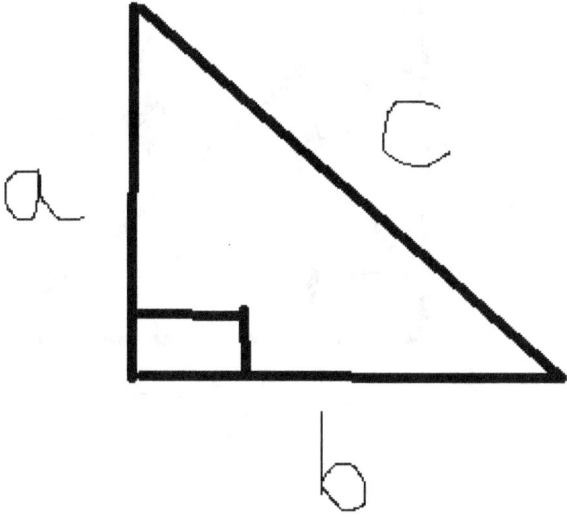

Simply put, the Pythagorean Theorem tells us that the sum of the squares of the legs equals the square of the hypotenuse. If we know two out of the three sides of a right triangle, we can find the third side by Pythagorean Theorem. The ordered triple (a, b, c) that is a solution to $a^2 + b^2 = c^2$ is called a Pythagorean Triple. The Pythagorean Theorem has many applications in several branches of mathematics, physics, engineering, astronomy, etc.

TYPES OF RIGHT TRIANGLES

The two types of right triangles every student should know are the 30-60-90 right triangle and the 45-45-90 right triangle. See the triangles on the next page.

Note: The sum of all three angles in any triangle in Euclidean Geometry (flat space) is 180 degrees.

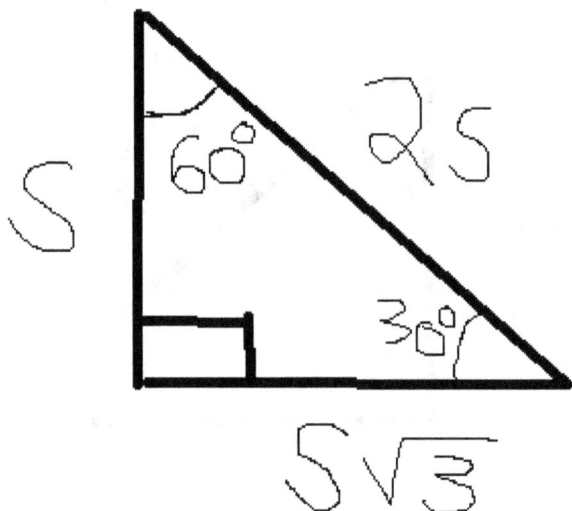

30-60-90 Right Triangle

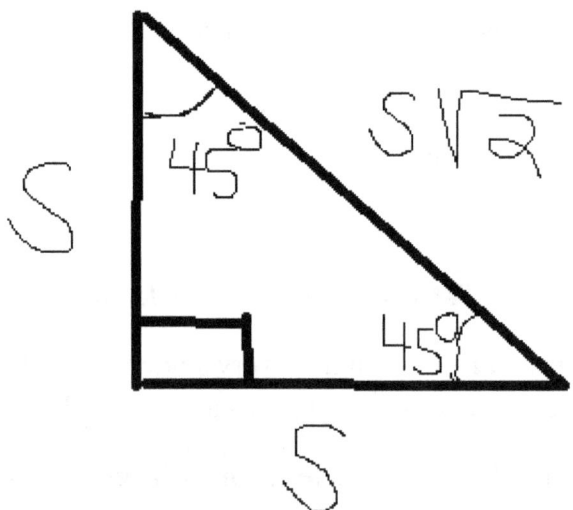

45-45-90 Right Triangle

32

One observation we notice in the 30-60-90 right triangle is the side with square root of 3 is adjacent to 30 degree angle and the hypotenuse is twice the remaining side opposite of the 30 degree angle. Also, we notice in the 45-45-90 right triangle is that both legs are the same length and the hypotenuse has a square root of 2. These are good facts to remember when you are dealing with these 2 types of right triangles.

OTHER TYPES OF TRIANGLES

Besides the right triangle, there are a few other types of triangles that the student should know.

EQUILATERAL/EQUIANGULAR: All three sides of this triangle are congruent. Their angles are congruent as well with each of them being 60 degrees.

ISOSCELES: At least two sides of this triangle are congruent. And at least two angles in this triangle are congruent.

SCALENE: All sides are different in this triangle as well as the angles.

ACUTE: This triangle has all acute angles, which are angles greater than 0 degrees and less than 90 degrees.

OBTUSE: This triangle has an obtuse angle, which is an angle greater than 90 degrees but less than 180 degrees.

SHAPES, SOLIDS, AND THEIR FORMULAS

There are important shapes that we have learned in geometry such as the square, rectangle, triangle, circle, etc. Since these shapes are two dimensional, we can calculate the perimeter (circumference for the circle) and area.

PERIMETER/CIRCUMFERENCE

SQUARE: $P = 4s$, where s is the side of the square.

RECTANGLE: $P = 2l + 2w$, where l is the length and w is the width of the rectangle.

TRIANGLE: $P = x + y + z$, where x, y, z are the three sides of any triangle.

CIRCLE: $C = \pi d$, where d is the diameter of the circle.

AREA

SQUARE: $A = s^2$

RECTANGLE: $A = lw$

TRIANGLE: $A = \frac{1}{2} bh$, where b is the base and h is the height.

CIRCLE: $A = \pi r^2$, where r is the radius of the circle.

Another object that was not on this list that is very important is the trapezoid. We are mostly interested in the area of the trapezoid.

TRAPEZOID: $A = \frac{1}{2}(B + b)h$, where B is the large base, b is the small base and h is the height.

There are important solids we have learned in geometry such as the cube, rectangular prism, sphere, cone, cylinder, etc. Since solids are three dimensional objects, we can calculate the surface area and volume of each of them.

SURFACE AREA

CUBE: $SA = 6x^2$, where x is the side of the cube.

RECTANGULAR PRISM: $SA = 2lw + 2lh + 2wh$

SPHERE: $SA = 4\pi r^2$

CONE: $SA = \pi r^2 + \pi ra$, where a is the slanted height of the cone.

CYLINDER: $SA = 2\pi r^2 + 2\pi rh$

VOLUME

CUBE: $V = x^3$

RECTANGULAR PRISM: $V = lwh$

SPHERE: $V = 4/3\pi r^3$

CONE: $V = 1/3\pi r^2 h$

CYLINDER: $V = \pi r^2 h$

Of course, there are many other shapes and solids we could list here. But these are the main ones the student will encounter frequently in advanced studies.

COMPUTATIONAL EXERCISES #3

1. Plot each of the following points and determine which quadrant or axis each point belongs to.

   a) $(2, 1)$

   b) $(-3, -5)$

   c) $(0, 4)$

   d) $(-1, 1)$

   e) $(7, 0)$

   f) $(8, -10)$

2. Find the distance between the points $(4, 6)$ and $(-1, -2)$.

3. Find the midpoint between the points $(-10, 8)$ and $(24, -50)$

4. Find a point on the y-axis that is equidistant from the points $(-3, 5)$ and $(4, 2)$.

5. Find the distance from the midpoint of $(a/2, -b)$ and $(-a, b/2)$ to $(a/4, b/4)$.

6. Find the perimeter/circumference and area of each of the following shapes.

   a) A square with a side length of 4 inches

   b) A rectangle with length of 12 feet and a width of 9 feet

   c) A right triangle with legs x, $x + 2$, and hypotenuse of $x + 4$ centimeters

   d) A semi-circle with a diameter of R meters

7. Find the surface area and volume of each of the following solids.

   a) A cube with a side of 7 inches

   b) A rectangular prism with a length of 15 feet, a width of 8 feet, and a height of 3 feet

   c) A sphere with a radius of $a/2$ centimeters

d) A cylinder with a diameter of d meters and the height h is twice the diameter.

CONCEPTUAL EXERCISES #3

1. True or False: If the x coordinate is negative and the y coordinate is positive, then the point $(x, -y)$ belongs in the 4th quadrant.

2. True or False: Any point that has a non-zero x value and a zero y value lies on the x-axis and can be written as $(x, 0)$.

3. True or False: The order in which you subtract the x and y coordinates in the distance formula does not matter since you arrive to the same result either way.

4. True or False: The midpoint is (x', y'), where x' is one half the difference between the x coordinates and y' is the average of the y coordinates.

5. True or False: The ordered triple (5, 12, 13) is a Pythagorean Triple.

6. True or False: Each angle formed by two adjacent sides in an octagon is 135 degrees.

7. True or False: The 30-60-90 right triangle is also an isosceles triangle.

8. True or False: The equilateral/equiangular triangle is also an isosceles triangle and an acute triangle.

9. True or False: The area of a circle is greater than its circumference for any radius.

10. True or False: If you double the width and height of a rectangular prism, then you increase its surface area by a factor of 8 and quadruple its volume.

11. If $A = (a, b)$, $B = (-b, -a)$, and C is the midpoint of segment AB, show that segments AC and BC are equal.

12. The coordinates (x, y), $(3x, y)$, $(x, 4y)$, and $(3x, 4y)$ form a quadrilateral.
A) Show that the quadrilateral is a rectangle.

B) Find a function $A(x)$ that describes the area of the rectangle in terms of the perimeter P.

C) Prove that the maximum area is proportional to the square of the perimeter, which is $A_{max} = cP^2$. What is the value of c?

13. Consider an equilateral triangle with all sides of length s. Show that the area of this triangle can be written as $A = \dfrac{\sqrt{3}}{4}s^2$. For what values of s is the area greater than the perimeter?

14. In order to make the ratio of the volume of a cone to the volume of a cylinder 1:1, by how many times must you increase the height of the cone?

15. A piece of wire with length L is to be cut in the form of a square and a semi-circle on top of the square. Both the length of the square and the diameter of the semi-circle are equal. Find an equation for the area of this structure in the form $A = kL^2$. What is the value of k?

16. Let a cube have a length of side x. For what values of x is the surface area greater than its volume? If the volume of a cube is V cubic units, then write a formula for the diagonal D in the form $D = n^p V^r$, where n is some positive integer and p and r are rational exponents.

17. If the number of sides of a polygon is even, show that the sum of all the interior angles is divisible by 360.

18. Prove that the sum of all the exterior angles of any polygon is 360 degrees.

19. Prove the Pythagorean Theorem.

20. Derive the formula for the area of the trapezoid.

REVIEW #4

SEQUENCES

In Review #2, we examined functions and their important properties. Now we will investigate a special type of function called a sequence. Most people usually think of sequences as a list of items or terms that hold a certain pattern or meaning. Even though this notion is actually true, we can mathematically define a sequence as a function whose domain is the natural numbers and the range is some subset of the real numbers (mostly integers and rational numbers). A fundamental difference we see between general, real valued functions and sequences is that real valued functions are continuous, lines and curves and sequences are discrete points that are not connected to one another. There are 4 types of sequences we want to look at in this section.

1. ARITHMETIC: An arithmetic sequence has a common difference between one term and its successive term.

2. GEOMETRIC: A geometric sequence has a common ratio between one term and its successive term.

3. ALTERNATING: An alternating sequence is a sequence where the signs alternate between terms.

4. RECURSIVE: A recursive sequence is a sequence where successive terms depend on the values of the previous terms.

5. MISCELLANEOUS: A sequence that is neither arithmetic nor geometric, but can be alternating and/or recursive.

The mathematical notation for the nth term of a sequence is a_n. Other letters such as b and c can be used to define sequences. We will see this later on in the calculus reviews. Below are the following equations used for arithmetic, geometric, alternating, and recursive.

ARITHMETIC: If a is the first term in an arithmetic sequence and d is the common difference between a term and its successor, then the nth term of this sequence can be defined as $a_n = a + (n - 1)d$.

GEOMETRIC: If a is the first term in a geometric sequence and r is the common ratio between a term and its successor, then the nth term of this sequence can be defined as $a_n = ar^{n-1}$.

ALTERNATING: If $f(n)$ is any function of the natural numbers, then an alternating sequence can be written as $a_n = (-1)^n f(n)$.

RECURSIVE:

Type 1: If k is any non-zero real constant, and an initial value a_1 is given, then a recursive sequence of type 1 can be written as $a_n = ka_{n-1}$, for any natural number $n \geq 2$.

Type 2: If c_1 and c_2 are any non-zero real constants, and two initial values a_1 and a_2 are given, then a recursive sequence of type 2 can be written as $a_n = c_1 a_{n-1} + c_2 a_{n-2}$, for any natural numbers $n \geq 3$.

Of course, the recursive sequence can be extended to Type N where we have N number of initial values. But for practical purposes, the type 1 and type 2 recursive sequences are the ones students will work with most frequently.

Miscellaneous sequences are the sequences that do not possess a common difference or common ratio patterns and can sometimes not have a specific formula at all. However, we have witnessed such sequences in calculus and will review their important properties later on in the text.

SERIES

A series is very similar to a sequence. Instead of listing all of the terms like we did in a sequence, a series gathers the sum of all the terms in a sequence. We can find the sum of all the terms of a finite, (not infinite) arithmetic series. But we can also find the sum of all the terms of both finite and infinite geometric series. Like the sequence, we also have a specific mathematical notation to use to define a series. We use the capital sigma, which is a Greek symbol meaning "sum." Below is how we can define both a finite and infinite series.

FINITE SERIES: Let a_n be any sequence. The finite series of a_n can be defined as follows:

$$\sum_{n=1}^{k} a_n = a_1 + a_2 + \ldots + a_k,$$ where k is some natural number greater than 1.

The letter n is known as the index of summation. The number 1 is the lower limit of the summation. The number k is the upper limit of the summation.

This tells us that we need to start our sum at $n = 1$ and all all of the terms until we reach $n = k$. Think of 1 as the starter and k as the stopper.

INFINITE SERIES: Let a_n be any sequence. The infinite series of a_n can be defined as follows:

$$\sum_{n=1}^{\infty} a_n = a_1 + a_2 + \ldots$$

Unlike the finite series where we have a countable number of terms to add up, the infinite series has an indefinite amount of terms to add. We may wonder if there exists a finite sum to the infinite series. This is a question that will be addressed in the near future and in the calculus review.

HOW TO FIND THE SUM OF A FINITE ARITHMETIC SERIES

We could take each of the terms of an arithmetic series one by one and add them up, but this would take up a great deal of time for a series that has 100s to 1000s of terms. There has to be a better way to find the sum of arithmetic series. The question is how do we find a formula that can give us the sum to any finite arithmetic series? Here is the derivation below.

Consider the formula for finding the nth term of an arithmetic sequence $a_n = a + (n-1)d$.

Let's list the first few terms of this sequence.

$a_1 = a$, $a_2 = a + d$, $a_3 = a + 2d$, $a_4 = a + 3d$

Now let S be the sum of all the finite terms in the arithmetic series.

$S = a + (a + d) + (a + 2d) + (a + 3d) + \ldots + (a + (n-1)d)$

We can also add the terms starting from the nth term all the back to the 1st term.

$S = (a + (n-1)d) + (a + (n-2)d) + (a + (n-3)d) + (a + (n-4)d) + \ldots + a$

If we add the two equations together, we get the following:

$2S = (2a + (n-1)d) + (2a + (n-1)d) + \ldots + (2a + (n-1)d)$, added n-times.

So, we have

$2S = n(2a + (n-1)d)$

Dividing by 2 to both sides, we have the formula for the sum.

$$S = \frac{n}{2}\left(2a + (n-1)d\right)$$

In order to find the finite sum of an arithmetic series, all we need is the total number of terms, the first term, and the common difference.

HOW TO FIND THE SUM OF A FINITE GEOMETRIC SERIES

Like the arithmetic series, we can also derive a formula for the sum of the finite geometric series.

Below is the derivation of the formula to find the sum of this series.

Consider the geometric sequence $a_n = ar^{n-1}$. We can list out the first few terms in this sequence.

$a_1 = a$, $a_2 = ar$, $a_3 = ar^2$, $a_4 = ar^3$

Let S be the sum of all the terms in the finite geometric series. Then we have the following:

$S = a + ar + ar^2 + ar^3 + \ldots + ar^{n-1}$

Now let's do an algebraic manipulation by multiplying r to S.

$rS = ar + ar^2 + ar^3 + \ldots + ar^{n-1} + ar^n$

If we subtract $S - rS$, we get

$S - rS = a - ar^n$. Notice that all of the terms between the first and last terms cancelled out. Now we factor out S on the left side and factor out a on the right side to get

$S(1 - r) = a(1 - r^n)$.

Dividing both sides by $1 - r$, we have the following formula for the sum of the finite geometric series.

$S = \dfrac{a(1 - r^n)}{1 - r}$, for all ratios $|r| < 1$, which translates to $-1 < r < 1$.

Therefore, we can find the sum of any finite geometric series as long as the common ratio r is any value in the interval $(-1, 1)$.

SUM FORMULA FOR AN INFINITE GEOMETRIC SERIES

The sum formula for an infinite geometric series can be found in a similar manner as its finite counterpart. The formula is almost identical as the previous one, but simpler.

$S = \dfrac{a}{1 - r}$, for all ratios $|r| < 1$.

Notice that the value of r has to lie in the interval $(-1, 1)$ to have a sum. Otherwise, the infinite geometric series has no sum.

MATHEMATICAL INDUCTION

One of the most widely used principles in all of mathematics is mathematical induction. This principle is used to prove many theorems that involve the natural numbers. Below is the general outline of mathematical induction

OUTLINE OF MATHEMATICAL INDUCTION

1. Let $P(n)$ be a statement for all the natural numbers n.

2. Check to see if $P(1)$ is true.

3. Assume that $P(k)$ is a true statement for all natural numbers k.

4. Prove $P(k + 1)$ is true. This is also known as the "inductive step."

5. Therefore, $P(n)$ is true for all natural numbers n.

A good real world situation that relates to the principle of mathematical induction is the falling of dominoes. We know from past experience that if you stack many dominoes in a straight row and knock the first domino down, then the next domino will fall down and so forth. Let us say that the first domino falls down is true and the nth domino falls down is also true. Then we can induce that the $n + 1$ domino falls down must be true as well.

The principle of mathematical induction is a powerful tool of inductive reasoning that helps us to prove statements for very large values of n. Proving cases for $n = 100$ or $n = 1,000$ does not validate truth for any statement. We have to show that the statement is true for all values of n.

FACTORIAL NOTATION AND THE BINOMIAL THEOREM

Suppose that we want to count numbers starting at some number n all the way back to the number 1 and multiply all of those numbers together. Then the following expression would look like

$n(n - 1)(n - 2)*...*3*2*1$

We can see that if n is very large, then this product will be very large as well. Mathematicians created a notation to describe these products, which are known as factorials.

For example, if we want to compute 3! (read as "3 factorial) then we would write

$3! = 3*2*1 = 6$. So, $3! = 6$.

$5! = 5*4*3*2*1 = 120$. So, $5! = 120$

$n! = n(n-1)(n-2)*...*3*2*1$ This is known as "n factorial."

Note: By definition $0! = 1$.

The factorial has many applications in counting, probability, statistics and many other branches of mathematics and the sciences. One specific formula we will look at is the general expression of a binomial to the nth power. This is known as the binomial theorem.

But first, we need to introduce some other mathematical tools to help us gain an understanding of the binomial theorem.

COMBINATION FORMULA

The combination formula serves both theoretical and practical purposes in mathematics. It is one of the main tools we need on our way to the binomial theorem. On a practical note, let us say you have n number of objects and you want to choose only r of those objects at a time. How many ways can we do this? The following formula gives us the answer to this mysterious question.

$$\binom{n}{r} = \frac{n!}{r!(n-r)!}, \text{ for } r \leq n$$

The notation on the left side of the equation is read as "n choose r."

This formula helps us to find the coefficients to each of the terms in the binomial.

Now that we have the combination formula under our belts, we can look at the binomials more closely. The question is how do we derive a formula for any binomial to the nth power?

We can organize our data in a table and observe some interesting patterns.

N	Binomial	Expansion	Coefficients
1	$(a+b)^1$	$a+b$	1, 1
2	$(a+b)^2$	$a^2 + 2ab + b^2$	1, 2, 1
3	$(a+b)^3$	$a^3 + 3a^2b + 3ab^2 + b^3$	1, 3, 3, 1
4	$(a+b)^4$	$a^4 + 4a^3b + 6a^2b^2 + 4ab^3 + b^4$	1, 4, 6, 4, 1

In these four cases, we see that the first and last coefficient of each binomial is always 1. We also observe that a binomial to the nth power has $n + 1$ coefficients. The first term a always starts with the nth power and decreases its power by 1 while the second term b appears to the first power and increases its power by 1 up to n. And the sum of the exponents of both terms a and b is always equal to n.

There is one question we should ask ourselves regarding the coefficients. Is there a way to gather the coefficients for the next power of n based on the previous coefficients? The answer is yes!

To get the coefficients, we can add adjacent coefficients to get the next coefficient. For example, to get the middle coefficient for the 2nd power binomial, we add 1 and 1 to get 2. To get the middle two coefficients for the 3rd power binomial, we add 1 and 2 to get 3 and 2 and 1 to get 3 again. That is why we have two 3's. Then we repeat this process to get the 4th power binomial and so forth. This structure is known as "Pascal's Triangle."

(In honor of the French mathematician Blaise Pascal)

$$
\begin{array}{ccccccccccc}
 & & & & & 1 & & & & & \\
 & & & & 1 & & 1 & & & & \\
 & & & 1 & & 2 & & 1 & & & \\
 & & 1 & & 3 & & 3 & & 1 & & \\
 & 1 & & 4 & & 6 & & 4 & & 1 & \\
1 & & 5 & & 10 & & 10 & & 5 & & 1
\end{array}
$$

And Pascal's Triangle goes on indefinitely for any natural number n.

Here is how the coefficients are written using the combination formula.

$$
1 = \binom{n}{0} = \binom{n}{n}, \ 2 = \binom{2}{1}, \ 3 = \binom{3}{1} = \binom{3}{2}, \ 4 = \binom{4}{1} = \binom{4}{3}, \ 5 = \binom{5}{1} = \binom{5}{4}
$$

Putting all this information together with the expansion of the binomials and their coefficients, we have the binomial theorem below.

BINOMIAL THEOREM: If n is any natural then the binomial $(a + b)^n$ can be expanded by the formula:

$$(a + b)^n = \binom{n}{0}a^n + \binom{n}{1}a^{n-1}b + \binom{n}{2}a^{n-2}b^2 + \ldots + \binom{n}{n}b^n$$

The binomial theorem can be proved by mathematical induction and applying the following combination sum formula. Sir Isaac Newton proved the binomial theorem circa 17th century.

COMBINATION SUM FORMULA: $\binom{n}{r-1} + \binom{n}{r} = \binom{n+1}{r}$

We may also be interested in finding any given term in the binomial expansion.

FINDING THE GENERAL TERM:

A term with that has a^r in expanding $(a + b)^n$ is the following:

$$\binom{n}{n-r}a^r b^{n-r}$$

This review has covered some abstract concepts of how to derive formulas involving sequences and series as well as the method of proving statements of the natural numbers with the principle of mathematical induction. We will revisit some of these concepts again in the calculus review section and unravel even more significant properties of sequences and series

COMPUTATIONAL EXERCISES #4

1. Find the first 4 terms of each of the following sequences. Determine whether each sequence is arithmetic, geometric, alternating, recursive, or neither.

   a) $a_n = 3n - 4$

   b) $a_n = 6(1/2)^{n-1}$

   c) $a_n = (-1)^{n+1}(n^2 + n - 1)$

   d) $a_n = \dfrac{5^n + 5}{n!}$

   e) $a_n = 2a_{n-1} - a_{n-2}$; $a_1 = 1$, $a_2 = 2$

2. Find a formula for an arithmetic sequence whose first term is -7 and common difference is 4. What is the 20th term of the sequence?

3. Find a formula for a geometric sequence whose second term is 54 and fourth term is 24. What is the 9th term of the sequence?

4. Find a formula for the recursive sequence $a_n = 3a_{n-1}$; $a_1 = 2$ Is the formula arithmetic or geometric?

5. Evaluate each of the following series.

   a) $\sum\limits_{n=1}^{3} 4n - 2$

   b) $\sum\limits_{i=0}^{5} (i+1)^2$

   c) $\sum\limits_{j=1}^{20} 5j + 10$

   d) $\sum\limits_{k=1}^{\infty} 8\left(\frac{7}{9}\right)^{k-1}$

   e) $\sum\limits_{m=1}^{6} (-1)^m \frac{m-2}{m^3}$

6. How many ways can one choose 4 members for a committee out of 10 potential candidates?

7. Simplify each of the factorial expressions.

   a) $n! / (n-2)!$

   b) $(2n+2)! / (2n)!$

   c) $(n-1)! / n! + (n+2)! / (n+1)!$

8. Expand each of the following binomials.

   a) $(x+2)^4$

   b) $(y-5)^3$

   c) $(2t+3)^6$

9. What is the 5th term in the binomial expansion of $(3z - 4)^8$

10. What is the coefficient of the 7th term in the binomial expansion of $(x + 2y)^{10}$?

CONCEPTUAL EXERCISES #4

1. True or False: The domain of a sequence is the set of natural numbers.

2. True or False: A sequence with a common difference between adjacent terms is an arithmetic sequence.

3. True or False: A geometric sequence always has a common denominator between adjacent terms.

4. True or False: It is possible to find the sum of an infinite arithmetic series.

5. True or False: The formula to find the sum of a finite geometric series is

$$S_n = \frac{a(1 - r^n)}{1 - r}, \text{ where } |r| < 1.$$

6. True or False: The formula to find the sum of a finite arithmetic series is

$$S_n = \frac{n}{2}(a + nd).$$

7. The combination formula $\binom{n}{r}$ is also equal to $\binom{n}{n - r}$.

8. The sum of the coefficients in the nth row of Pascal's triangle is 2^n.

9. Let a_n and b_n be arithmetic sequences and c_n and d_n be geometric sequences.

   a) Show that $a_n + b_n$ is an arithmetic sequence.

   b) Show that $c_n d_n$ is a geometric sequence.

10. Show that $0.999\ldots = 1$.

11. Show that the decimal number in the form $a.bbb\dots$ can be written as a rational number. If a and b are even integers, show that the rational number can be written in the form $2t/9$, where t is some integer.

12. A bouncy ball is held up in the air at a certain height h and is released. After each bounce, the ball's height is reduced by 25%. What is the vertical distance traveled by the ball in terms of h after an indefinite number of bounces?

13. Prove each of the following summation properties. Let c be a real constant.

a) $\displaystyle\sum_{n=1}^{k}(a_n + b_n) = \sum_{n=1}^{k}a_n + \sum_{n=1}^{k}b_n$

b) $\displaystyle\sum_{n=1}^{k}ca_n = c\sum_{n=1}^{k}a_n$

c) $\displaystyle\sum_{n=1}^{k}c = kc$

14. Prove the combination sum formula. Hint: Get a common denominator and combine the numerators and simplify.

15. Prove each of the following statements by the principle of mathematical induction.

a) $1 + 2 + 3 + \dots + n = n(n+1)/2$

b) $1 + 3 + 5 + \dots + 2n - 1 = n^2$

c) $a + ar + ar^2 + \dots + ar^{n-1} = \dfrac{a(1 - r^n)}{1 - r}$

d) $7^n - 4^n$ is divisible by 3.

e) $2^n \leq n!$ for all $n \geq 4$.

f) Binomial Theorem Hint: Use the combination sum formula in the inductive step.

REVIEW #5

MATRICES

In college and higher level mathematics, matrices are some of the most useful tools to solving complex problems. Matrices help us better understand the study of linear systems such as determining whether a linear system has a solution and how many solutions exist. They even help us in understanding geometric structures in the physical sciences. Without matrices, some of the technological advances we enjoy in our daily lives such as HD television, computers, cell phones, GPS, and video games would not exist.

Let us review what a matrix is and some of its unique properties.

MATRIX: A matrix is a rectangular (or square) array that can store numbers, functions, and other objects. A matrix has a certain number of rows and columns. An entry is stored in a particular row and column of the matrix. Here is how we mathematically define a matrix.

Let A be an m x n (read as "m by n") matrix. This is the dimension of the matrix A, which has m number of rows and n number of columns, where $m \neq n$. The entry that is stored in the ith row and jth column is denoted as a_{ij}.

For example, if we have A be a 2 x 3 matrix, then the matrix has 2 rows and 3 columns. If the entry $a_{12} = 4$, then the number 4 belongs to the 1st row and 2nd column location of matrix A.

SQUARE MATRIX: A square matrix A is an n x n matrix. This means that the matrix A has the same number of rows as columns.

Both rectangular and square matrices serve important roles in the study of linear algebra and possess their individual properties. Now we will look at the operations we can do on matrices.

OPERATIONS ON MATRICES

Out of the 4 operators, we can add, subtract, and multiply two or more matrices. Division is not defined with matrices. Here are some rules we need to know about matrices in order to know whether a certain operation is possible.

Let A and B be matrices

ADDITION: In order to add $A + B$, both matrices A and B have to be the same dimension. To get the new entries for $A + B$, we add their corresponding entries together.

SUBTRACTION: The rule of addition also applies for subtracting $A - B$, which is the same as $A + (-B)$. This means we take the opposite sign of all the entries of B and add their corresponding entries to A.

MULTIPLICATION: A x B is possible if and only if the number of columns in A is equal to the number of rows in B. That is, if A is an m x n matrix and B is a n x p matrix, then A x B is possible. The new matrix A x B will have a dimension of m x p. In other words, the A x B matrix will have the same number of rows in A and the same number of columns in B.

The way we find the any ij entry in A x B is to take all of the entries in row i of A and multiply it to all the corresponding entries in column j of B and then add the products. This is known as a "dot product", which is very useful when we talk about vectors in physics.

OPERATIONS ON A SINGLE MATRIX

There are 3 possible operations we can perform on a single matrix. These operations will play an important role when solving linear systems and finding the inverse of a matrix.

1. Multiply a row by a non-zero scalar: We can take any non-zero real number and multiply it to all of the entries in a certain row of the matrix. This is denoted as cR_k, where c is any non-zero real number and R_k is the kth row of the matrix.

2. Switch Rows: We can switch the entries from one row to another. This is denoted as

    $R_k \longleftrightarrow R_m$. This means all of the entries in the kth row go to the mth row and vice versa.

3. Add one row to another: We can add all of the entries in one row to their corresponding entries in another row to get a new row in the matrix.

    This is denoted as $R_k + R_m$. It is important to note that we can also multiply a scalar to the kth row and add those entries to the mth row. The mth row gets the new entries while the entries in the kth row do not change.

DETERMINANTS

A special property of square matrices is something known as "determinants." Determinants are helpful in solving systems of linear equations as well as finding the inverse of a square matrix. It is important to note that only square matrices with dimension n x n have determinants. Rectangular matrices with m x n dimension do not apply. Let us review the definition of a determinant for a 2 x 2 and a 3 x 3 matrix.

DETERMINANT OF 2 x 2 MATRIX: Let $A = \begin{bmatrix} a_{11} & a_{12} \\ a_{21} & a_{22} \end{bmatrix}$ be a 2 x 2 matrix. Then the determinant of A is defined as $\det(A) = a_{11}a_{22} - a_{12}a_{21}$.

Note: Some mathematicians like to denote det(A) as $|A|$, which is not to be confused with the absolute value of A.

All we do when taking a determinant of a 2 x 2 matrix is take the cross product and subtract. The 3 x 3 determinant is a little more complicated, but can be found with careful computation.

DETERMINANT OF 3 x 3 MATRIX: Let $A = \begin{bmatrix} a_{11} & a_{12} & a_{13} \\ a_{21} & a_{22} & a_{23} \\ a_{31} & a_{32} & a_{33} \end{bmatrix}$ be a 3 x 3 matrix. Then the

determinant of A is defined as

$$\det(A) = a_{11}(a_{22}a_{33} - a_{23}a_{32}) - a_{12}(a_{21}a_{33} - a_{23}a_{31}) + a_{13}(a_{21}a_{32} - a_{22}a_{31}).$$

We can break down this formula into 3 parts. First, we take the a_{11} entry and multiply that by the determinant of a sub-matrix of A. The sub-matrix is found by covering the first row and first column of the original matrix. Second, we take the opposite sign of the a_{12} entry and multiply that by the determinant of another sub-matrix of A. This sub-matrix is found by covering the first row and second column of the original matrix. Finally, we take the a_{13} entry and multiply this to a third sub-matrix of A. This sub-matrix is found by covering the first row and third column of the original matrix. Then we add all of these results together to obtain the determinant of a 3 x 3 matrix.

Another way to visualize (and perhaps memorize) the 3 x 3 determinant is with this formula.

$$\det(A) = a_{11}\begin{vmatrix} a_{22} & a_{23} \\ a_{32} & a_{33} \end{vmatrix} - a_{12}\begin{vmatrix} a_{21} & a_{23} \\ a_{31} & a_{33} \end{vmatrix} + a_{13}\begin{vmatrix} a_{21} & a_{22} \\ a_{31} & a_{32} \end{vmatrix}$$

This version of the 3 x 3 determinant formula gives us the actual method on finding the determinant with less letters and subscripts to memorize.

THE INVERSE OF A MATRIX

In algebra, we learned that numbers have inverses under addition and multiplication. For example, the additive inverse of a real number a is $-a$. If we add the number a and its additive inverse we get the identity, which is 0. Another example is the multiplicative inverse of a real number a is $1/a$. If we multiply the number and its multiplicative inverse we get the identity, which is 1. Unfortunately, not all real numbers have a multiplicative inverse. For instance, 0 does not have a multiplicative inverse since $1/0$ is undefined.

The same is true for square matrices. Some square matrices have an inverse while others do not. The question is how do we find the inverse of a matrix and how do we know whether the inverse of a matrix exists? The second part of the question is very easy to answer, which deals with the concept of determinants we covered earlier.

Note: The inverse of a square matrix A is denoted as A^{-1}.

EXISTENCE OF INVERSE: Let A be a n x n matrix. A^{-1} exists if and only if det(A) is non-zero.

In other words, we know that a matrix with a zero determinant cannot have an inverse.

It is important to know that the inverse of a matrix is unique.

A matrix A multiplied by it inverse A^{-1} yields the identity matrix I. That is, $AA^{-1} = A^{-1}A = I$. The identity matrix is a square matrix that contains 1's on the main diagonal and 0's everywhere else.

But how do we find A^{-1}? The process involves combining the 3 possible row operations of a matrix that we discussed earlier. We can form a special matrix $[A \mid I]$ known as the augmented coefficient matrix. The goal is to find A^{-1} so that we have the resulting augmented coefficient matrix $[I \mid A^{-1}]$. Notice that the original matrix starts on the left side and the identity matrix starts on the right side. After we do some row operations we get the identity matrix on the left side and the inverse matrix is on the right side.

Below is the formula for finding the inverse of a 2 x 2 matrix.

$$A^{-1} = \frac{1}{\det(A)} \begin{bmatrix} a_{22} & -a_{12} \\ -a_{21} & a_{11} \end{bmatrix}$$

All we do to find the inverse matrix is find the determinant of the original matrix, switch the entries on the main diagonal, and negate the entries on the reverse diagonal. Then we multiply the reciprocal of the determinant to each of the entries in the inverse matrix.

In order to find the inverse of a 3 x 3 matrix, we need to understand another property of matrices known as the "transpose."

TRANSPOSE OF A MATRIX

We may have heard the word "transpose" in other subjects such as chess and music where something develops from one form into another. The transpose of a matrix is similar. The transpose of A denoted as A^{T} tells us to interchange the entries of the rows and columns in the original matrix. That is, if the ij-entry of A is a_{ij}, then that entry in A^{T} is the ji-entry or a_{ji}. Here are the transposes of the general 2 x 2 and 3 x 3 matrices.

TRANSPOSE OF 2 x 2: $A^T = \begin{bmatrix} a_{11} & a_{21} \\ a_{12} & a_{22} \end{bmatrix}$

TRANSPOSE OF 3 x 3: $A^T = \begin{bmatrix} a_{11} & a_{21} & a_{31} \\ a_{12} & a_{22} & a_{32} \\ a_{13} & a_{23} & a_{33} \end{bmatrix}$

Now we can move forward in determining the inverse of a 3 x 3 matrix.

INVERSE OF A 3 x 3 MATRIX

Finding the inverse of a 3 x 3 matrix can be a little cumbersome and tedious, but we will try to simplify the steps to make computation easier. Let us follow the steps below.

HOW TO FIND THE INVERSE OF A

1. Compute det(A).

2. Find A^T.

3. Find the co-factors of A^T. What are the co-factors?

    To find each co-factor of A^T, we cover up the i-row and j-column of A^T and compute the determinant of the 2 x 2 sub-matrix. The sign of the co-factor entries c_{ij} depends on the formula sign = $(-1)^{i+j}$.

4. Place all 9 co-factors in their appropriate entries to form a new matrix known as the adjoint matrix or Adj(A).

5. Finally, we use the inverse formula. $A^{-1} = 1/\det(A)*\text{Adj}(A)$

Notice the similarity between calculating the 2 x 2 and 3 x 3 inverses. Both of them have the reciprocal of the determinant, which is the scalar multiplied to each of the entries in the new matrix.

SOLVING LINEAR SYSTEMS WITH MATRICES

Earlier in this book, we reviewed how to solve systems of linear equations by graphing, substitution, and elimination. We can also solve a linear system of equations by writing the system in matrix form. Consider the general 2 x 2 system of linear equations below.

$ax + by = e$

$cx + dy = f$

The matrix form can also be written as $As = b$, where $A = \begin{bmatrix} a & b \\ c & d \end{bmatrix}$, $s = \begin{bmatrix} x \\ y \end{bmatrix}$, and $b = \begin{bmatrix} e \\ f \end{bmatrix}$.

Notice that A is a 2 x 2 and s is a 2 x 1 matrix. So, when we multiply these together, we get b, which is a 2 x 1 matrix. Matrices with n rows and only one column are known as column vectors. (Row vectors are matrices with 1 row and n columns).

The question is how do we solve this system of linear equations in matrix form? The answer is very easy. Think about how we solved the equation $ax = b$ for x. We divided a to both sides to get $x = b/a$. We did the inverse operation of multiplication, which was division to solve for x. The same principle applies here with solving linear systems in matrix form. We have to take the inverse matrix A^{-1} of both sides to solve for s.

Therefore, the solution to our system is $s = A^{-1}b$. We can apply this procedure to solving 3 x 3 systems and up to n x n systems. However, the computations by hand can take quite a bit of time and may require a calculator or computer to perform accurate and precise computations.

We can also solve linear systems by method of determinants. This is known as "Cramer's Rule."

CRAMER'S RULE

Cramer's Rule is an efficient method for solving linear systems by determinants. The rule for solving 2 x 2 systems by Cramer's Rule is simple. Consider the following matrices.

$$A = \begin{bmatrix} a & b \\ c & d \end{bmatrix}, M = \begin{bmatrix} e & b \\ f & d \end{bmatrix}, \text{ and } N = \begin{bmatrix} a & e \\ c & f \end{bmatrix}$$

Then $x = \det(M) / \det(A)$ and $y = \det(N) / \det(A)$.

Cramer's Rule for 3 x 3 systems is similar, but a little more complicated. Consider the following matrices below.

$$A = \begin{bmatrix} a & b & c \\ d & e & f \\ g & h & i \end{bmatrix}, R = \begin{bmatrix} j & b & c \\ k & e & f \\ l & h & i \end{bmatrix}, S = \begin{bmatrix} a & j & c \\ d & k & f \\ g & l & i \end{bmatrix}, \text{ and } T = \begin{bmatrix} a & b & j \\ d & e & k \\ g & h & l \end{bmatrix}$$

Then $x = \det(R) / \det(A)$, $y = \det(S) / \det(A)$, and $z = \det(T) / \det(A)$.

Note that Cramer's Rule can only be applied for linear systems that have n equations and n unknowns. Also, $\det(A) \neq 0$.

GAUSSIAN ELIMINATION

Gaussian Elimination is a great method to solving any system of linear equations. It is mostly used for solving 3 x 3 systems and higher. Recall the following system below.

$ax + by + cz = j$

$dx + ey + fz = k$

$gx + hy + iz = l$

This system can be written as the following augmented coefficient matrix.

$$\begin{bmatrix} a & b & c & j \\ d & e & f & k \\ g & h & i & l \end{bmatrix}$$

The goal in Gaussian Elimination is to use any of the 3 row operations to convert the 3 x 3 matrix in the following form.

$$\begin{bmatrix} c_1 & c_2 & c_3 & A \\ 0 & c_4 & c_5 & B \\ 0 & 0 & c_6 & C \end{bmatrix}$$

Notice the arrangement of the zeros in triangular form. Now we have reduced the original system to the following equations. This process is known as "row reduction."

1. $c_1x + c_2y + c_3z = A$

2. $c_4y + c_5z = B$

3. $c_6z = C$

Now we solve for z in equation 3 and then plug the value of z in equation 2 and solve for y. Then we solve for x by plugging the values of y and z into equation 1. This process is known as "back substitution."

Gaussian elimination tells us to get a zero or eliminate the x variable in the second row and then get two zeros by eliminating the x and y variables in the third row. Then we work backwards to get the solution.

This brings up an important question. What happens if we get all zeros on the last row or all zeros on the last row except the final entry? Here are a couple of cases to consider.

ALL ZEROS CASE: If an $n \times n$ matrix has all zeros in the last row after row reduction, then the system has infinitely many solutions.

ALL ZEROS BUT LAST ENTRY CASE: If an $n \times n$ matrix has all zeros except the last entry in the final row after row reduction, then the system has no solution.

Gaussian Elimination was invented by the famous German mathematician Carl F. Gauss circa 19th century. Besides mathematics, Gauss contributed to many other scientific fields such as astronomy, physics, and engineering. In his homeland, Gauss is hailed as one of the mathematical giants in his day.

PARTIAL FRACTIONS DECOMPOSITION OF RATIONAL FUNCTIONS

In this topic, it appears that we are taking a slight detour from the subject of matrices. However, we will see that the topic of partial fractions is an application for matrices, especially for 3×3 linear systems and higher.

Recall that a rational function $r(x)$ can be written in the form $r(x) = p(x) / q(x)$, where $p(x)$ are $q(x)$ are polynomial functions and $q(x) \neq 0$. The question is can we rewrite $r(x)$ as a sum of individual functions or partial fractions? The answer is yes! But we must consider a few cases for the denominator $q(x)$ in order to know how to break up the rational function as a sum of partial fractions.

CASE 1: $q(x)$ is written as a product of distinct linear factors.

This means that if $q(x) = (a_1 x + b_1)(a_2 x + b_2)\ldots(a_n x + b_n)$ then $r(x)$ can be written as

$$\frac{C_1}{a_1 x + b_1} + \frac{C_2}{ax_2 + b_2} + \ldots + \frac{C_n}{a_n x + b_n}.$$

CASE 2: $q(x)$ is written as a linear factor to the nth power.

This means if $q(x) = (ax + b)^n$ then $r(x)$ can be written as

$$\frac{C_1}{ax + b} + \frac{C_2}{(ax + b)^2} + \ldots + \frac{C_n}{(ax + b)^n}$$

CASE 3: $q(x)$ is written as a product of irreducible quadratic factors.

Note: An irreducible quadratic is a quadratic that cannot be factored as a product of linear factors, which does not contain any real roots over the field of real numbers.

This means if $q(x) = (a_1x^2 + b_1)(a_2x^2 + b_2)\ldots(a_nx^2 + b_n)$ then $r(x)$ can be written as

$$\frac{C_1x + D_1}{a_1x^2 + b_1} + \frac{C_2x + D_2}{a_2x^2 + b_2} + \ldots + \frac{C_nx + D_n}{a_nx^2 + b_n}.$$

Note that irreducible quadratic expressions can also be trinomials.

CASE 4: $q(x)$ is written as an irreducible quadratic factor to the nth power.

This means if $q(x) = (ax^2 + b)^n$ then $r(x)$ can be written in the form of the partial fraction like CASE 3 and extended to n fractions with the same factor to the nth power like CASE 2.

A miscellaneous case is when we have both linear and irreducible quadratic factors. In this case, we use a conjunction of two or more cases that we stated above.

Now that we have all of the possible cases spelled out, the question is how do we find the constants in the numerators? Here are a few steps on the next page that can help us get started.

HOW TO FIND THE CONSTANTS IN THE NUMERATOR

1. Find the least common denominator (LCD).

2. Multiply the LCD to both sides of the equation. This will clear the fraction and leave only $p(x)$ on the left hand side.

   After step 2, there are two methods we can use to solve for the constants in the numerator.

   COEFFICIENT EQUIVALENCE METHOD:

   1. Distribute all of the constants to each term.

   2. Combine like terms.

   3. Set the sum of the constants on the right side of the equation for a certain term equal to the coefficient of its like term on the left side of the equation.

   4. Repeat step 3 until all equations have been found.

5. Solve the linear system of equations by substitution, elimination, or matrix method.

The other method does not involve distributing the constants to all of the terms, but allows us to keep everything in factored form.

CHOOSE AND CANCEL METHOD:

1. Pick a value for x that will cancel all of the constants but one.

2. Plug in the value of x on both sides of the equation.

3. Solve for the remaining unknown constant.

4. Repeat steps 1, 2, and 3 until all constants have been found.

Note in step 1 that we may not always eliminate all constants but one. Sometimes, we may have 2 or more constants left over in the equation, and thus have a system of linear equations to solve. Ideally, we want to eliminate as many constants as possible.

Both methods have their advantages and drawbacks. The method you prefer to utilize depends on how many constants you need to find and the LCD.

Partial fractions have other applications in higher level mathematics such as evaluating integrals in calculus. It helps to break up a rational function into smaller pieces and integrate part by part. We will see this in the calculus review.

COMPUTATIONAL EXERCISES #5

1. Determine the dimension of each matrix and categorize the matrix as rectangular or square. List all of the ij-entries of each matrix.

a) $A = \begin{bmatrix} 2 & -1 & 0 \\ -3 & 4 & 6 \end{bmatrix}$

b) $B = \begin{bmatrix} 1 & 7 \\ 5 & 9 \end{bmatrix}$

c) $C = \begin{bmatrix} 1 \\ 0 \\ -8 \end{bmatrix}$

d) $D = \begin{bmatrix} -2 & 5 & 10 \\ 9 & -3 & 27 \\ -18 & -15 & 6 \end{bmatrix}$

2. Consider the following matrices. $A = \begin{bmatrix} 1 & 2 \\ 3 & 4 \end{bmatrix}$, $B = \begin{bmatrix} -5 & 9 & 12 \\ 6 & -4 & 7 \end{bmatrix}$, $C = \begin{bmatrix} -10 & 20 \\ 16 & -8 \end{bmatrix}$.

Compute all of the following operations (if possible).

a) $A + C$

b) $C - 2A$

c) AB

d) $3A + 1/2C$

e) $B^T C$

f) $5B$

g) CB

h) AC

i) CA^T

j) $(AC)A$

3. Suppose $A = \begin{bmatrix} a & 0 & 0 \\ -a & 1 & 0 \\ a & 0 & -1 \end{bmatrix}$, where a is a non-zero real number. Compute each of the

following row operations. What kind of matrix do you get as a result?

a) $R_1 + R_2$

b) $-R_1 + R_3$

c) $1/a\, R_1$

d) $-R_3$

4. Calculate the determinant of each matrix.

a) $A = \begin{bmatrix} 2 & 4 \\ 1 & 5 \end{bmatrix}$

b) $B = \begin{bmatrix} -3 & 0 & 1 \\ 0 & 6 & -2 \\ -1 & 9 & 7 \end{bmatrix}$

5. A square matrix A is symmetric if and only if $A = A^T$. Determine whether each of the following matrices is symmetric.

a) $\begin{bmatrix} 1 & 2 \\ 2 & 1 \end{bmatrix}$

b) $\begin{bmatrix} 3 & 0 & 4 \\ 0 & 7 & 5 \\ 4 & 5 & 8 \end{bmatrix}$

c) $\begin{bmatrix} -6 & 12 & 9 & -6 \\ 12 & 0 & -4 & 12 \\ 9 & -4 & 8 & 9 \\ -6 & 12 & 9 & -6 \end{bmatrix}$

6. Find the unknown value(s) that guarantees the matrix is non-invertible (no inverse exists).

a) $\begin{bmatrix} x & 4 \\ 8 & 2 \end{bmatrix}$

b) $\begin{bmatrix} x & 2 & -1 \\ -1 & x & 1 \\ x & -3 & 0 \end{bmatrix}$

7. Determine whether the following matrices have inverses. Find the inverse of the matrices that have one.

a) $\begin{bmatrix} 14 & 7 \\ 4 & 2 \end{bmatrix}$

b) $\begin{bmatrix} -3 & 1 \\ 2 & -6 \end{bmatrix}$

c) $\begin{bmatrix} 9 & 0 & 3 \\ -4 & 1 & 0 \\ 0 & -2 & 5 \end{bmatrix}$

d) $\begin{bmatrix} 7 & 0 & 1 \\ 6 & 0 & 2 \\ 5 & 0 & 3 \end{bmatrix}$

8. Use Cramer's Rule to solve each of the following systems.

a) $\begin{array}{l} 2x + y = 1 \\ 3x - 4y = 18 \end{array}$

b) $\begin{array}{l} x + y + z = 2 \\ x + 2y - z = 6 \\ 2x - y + 5z = -5 \end{array}$

9. Solve each of the following systems by Gaussian Elimination.

a) $\begin{array}{l} -x + 3y = 9 \\ 6x + 2y = -12 \end{array}$

b) $\begin{array}{l} 2x - y + 4z = 10 \\ x + y - 2z = -4 \\ -2x + 3y + z = 1 \end{array}$

10. Write each of the rational functions as a sum of partial fractions.

a) $\dfrac{2x + 8}{x^2 - 4}$

b) $\dfrac{x^2 + x + 1}{x^3 + x^2 - x - 1}$

c) $\dfrac{6 - x}{x^3 + 64}$

CONCEPTUAL EXERCISES #5

1. True or False: The difference between a rectangular matrix and a square matrix is a rectangular matrix has unequal number of rows and columns while the square matrix has the same number of rows as columns.

2. True or False: It is possible to add and subtract matrices that do not have the same dimension.

3. True or False: In order to multiply two matrices, the first matrix must have the same number of columns as rows in the second matrix.

4. True or False: Any matrix with a zero determinant has no inverse.

5. True or False: When you multiply the matrix and its inverse in either order, you get the identity matrix. The identity matrix is a matrix that has 1's on the main diagonal and 0's everywhere else.

6. True or False: The transpose of a matrix is when you interchange the row and column entries of a matrix.

7. True or False: If all zeros exist on the last row after you row reduce a matrix, then you have only one solution.

8. True or False: The final step in the process of Gaussian Elimination is back substitution.

9. True or False: The solution to the system $As = b$ is $s = b / A^{-1}$.

10. True or False: A rational function with n factors in the denominator can be broken up as a sum of n partial fractions.

11. Determine whether each of the following statements of matrices is true or false. If it is true, verify it with general 2 x 2 matrices. If it is false, provide a counterexample to show why it is false. Then write an accurate statement to make it true.

    a) $A + B = B + A$

    b) $AB = BA$

    c) $A(BC) = (AB)C$

    d) $(A + B)^T = A^T + B^T$

e) $(AB)^T = B^T A^T$

12. Show that $\det(AB) = \det(A)\det(B)$ using 2 x 2 matrices. Is this true for any n x n matrix?

13. Prove that if a 2 x 2 matrix A is symmetric, then $\det(A) = \det(A^T)$. Is the converse of this statement true? If it is false, give a counterexample to show why it is false.

14. Verify the inverse property of $(AB)^{-1} = B^{-1}A^{-1}$ if $A = \begin{bmatrix} 3 & 1 \\ 2 & -1 \end{bmatrix}$ and B

$= \begin{bmatrix} 1 & 0 \\ -4 & 6 \end{bmatrix}$

15. If $B = A^{-1}$ and $C = B^{-1}$, then show that $A = C$.

16. The trace of a square matrix A denoted as Tr(A) is the sum of all the diagonal elements. Using the summation properties in the last review section, show that $Tr(A + B) = Tr(A) + Tr(B)$. Is this true for multiplication?

17. Consider the 2 x 2 system.

$$ax + by = a$$
$$bx - ay = b$$

Find the values of a and b that allow the system to have infinitely many solutions. Are they real or complex values?

18. Use the matrix $A = \begin{bmatrix} a & b \\ -b & a \end{bmatrix}$ to show that $(A^{-1})^T = (A^T)^{-1}$

19. Show that $\dfrac{a}{x(x^2 + a^2)} = \dfrac{1}{ax} - \dfrac{x}{(\sqrt{ax})^2 + a^4}$.

REVIEW #6

CONIC SECTIONS

We are reverting back to an important topic in geometry that provides us with an analysis of one of the most beautiful objects in mathematics, i.e. the cone. The ubiquitous cone has appeared in many areas of the real world such as theater, machinery, and ice cream. If we take the time and effort to cut up the cone into various cross-sections, we will notice different shapes arise in the matter. All of these shapes belong to a collection of objects known as "conic sections." Let us examine the 4 different types of conic sections: parabola, circle, ellipse, and hyperbola.

THE PARABOLA

If we hold a cone upside down where the base is on the bottom and cut the cone at an angle, we obtain the conic section known as the "parabola." A parabola is mathematically defined as the set of all points in a plane that is equidistant from a point known as the focus and a line called the directrix. The point where the parabola opens is called the vertex. The focus lies in front of the vertex while the directrix lies behind the vertex. The distance from the vertex to the focus is equal to the distance from the vertex to the directrix. The axis of symmetry is a line perpendicular to the directrix that cuts the parabola into equal halves. Parabolas can be categorized as either vertical or horizontal. Vertical parabolas can open upward or downward. Horizontal parabolas can open left or right. Parabolas are highly prevalent in designing headlights, telescopes, satellite dishes, etc. See the figure below of a vertical parabola on the next page. The vertex, focus, directrix, and axis of symmetry are all labeled.

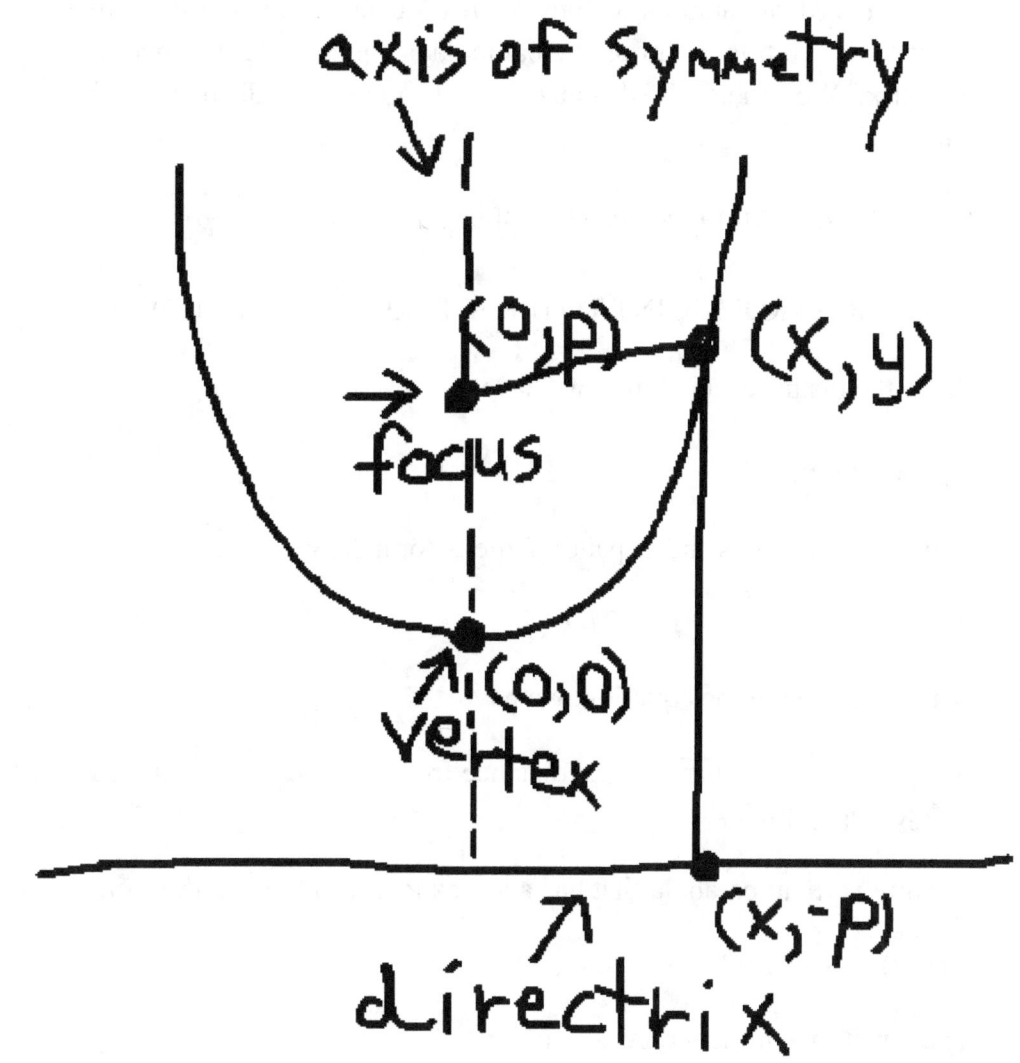

This is a vertical parabola with the vertex located at the origin. Notice the distance from the focus to an arbitrary point on the parabola is equal to the distance from the arbitrary point to the directrix. We can use the distance formula and set the distances equal to find an equation for the parabola.

DISTANCE FROM FOCUS TO POINT: $d_1 = \sqrt{x^2 + (y - p)^2}$

DISTANCE FROM POINT TO DIRECTERIX: $d_2 = \sqrt{(y + p)^2} = y + p$

Since the distances are equal, we have

$$\sqrt{x^2 + (y - p)^2} = y + p.$$

Squaring both sides and expanding the binomials, we have

$$x^2 + y^2 - 2py + p^2 = y^2 + 2py + p^2.$$

After simplifying the equation, we have

$x^2 = 4py$. Notice that p is the distance from the vertex to the focus, and the distance from the vertex to the directrix.

For any vertical parabola that has a vertex located at (h, k), this equation can be modified to

$$(x - h)^2 = 4p(y - k).$$

The equation for the directrix is $y = -p$.

The equation for the vertical axis of symmetry is $x = h$.

If we use the same procedure from the figure of the vertical parabola, then we can find an equation for the horizontal parabola whose vertex is at the origin.

$y^2 = 4px$.

Notice that we switched the x and y variables in the previous equation.

For any horizontal parabola that has a vertex located at (h, k), this equation can be modified to

$(y - k)^2 = 4p(x - h)$.

The equation for the directrix is $x = -p$.

The equation for the horizontal axis of symmetry is $y = k$.

Another important feature of the parabola is something known as the "latus rectum." This term is Latin for "straight side." The latus rectum is a chord that passes through the focus, which is parallel to the directrix. The length L of the latus rectum is $L = 4p$, which is the term that we see in both the vertical and horizontal parabolic equations.

THE CIRCLE

If we slice the cone above and parallel to its base we get the conic section known as the "circle." A circle is mathematically defined as the set of all points in the plane that are equidistant from a fixed point called the center. The distance from the center to any point located on the circle is called the radius. The diameter is a line segment that passes through the center of the circle and its endpoints are located on the circle. The diameter's length is twice the length of the radius. The circle is a widely used shape in geometry as well as trigonometry and calculus. See the figure below of a circle.

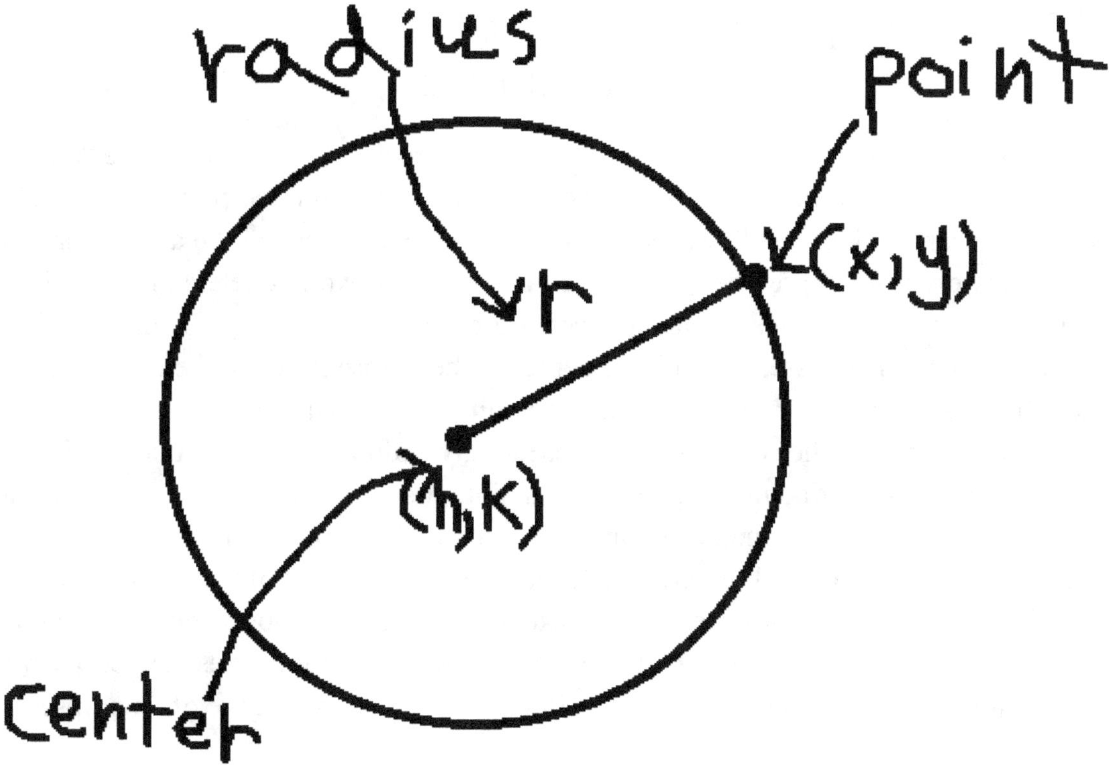

The question is can we find an equation of a circle that relates any point (x, y) to the radius r? The answer is yes! We use the distance formula to derive the equation of a circle.

Let (h, k) be the center of the circle, (x, y) be any point on the circle, and r be the radius of the circle.

By applying the distance formula, we have

$$(x - h)^2 + (y - k)^2 = r^2.$$

When expanding the binomials on the left side, we have

$$x^2 - 2hx + h^2 + y^2 - 2ky + k^2 = r^2$$

OR

$$x^2 + Bx + C + y^2 + Dy + E = r^2, \text{ where } B = -2h, C = h^2, D = -2k, \text{ and } E = k^2.$$

This is known as the expanded form of a circle equation, which is a quadratic equation in two variables. We can convert the expanded form to regular form by means of completing the square with the x and y terms.

THE ELLIPSE

If we slice the cone at an acute angle, then we get the conic section known as the "ellipse." An ellipse is the set of all points of the plane whose distances to two fixed points sum up to a constant. We can think of an ellipse as a closed and bounded curve. Like a circle, an ellipse has a center. However, an ellipse does not have a radius. It has two diameters of axes. The maximum diameter that passes through the center of the ellipse is called the major axis. The minimum diameter that passes through the center of the ellipse is called the minor axis. The segment from the center of the ellipse to any endpoint on the major axis is called the semi-major axis. The segment from the center of the ellipse to any endpoint on the minor axis is called the semi-minor axis. The endpoints that lie on the major axis are called the vertices. The two fixed points that are located on the major axis are called the foci (each fixed point is a focus). An ellipse can be either horizontal or vertical, depending on the length of the horizontal and vertical axes. Ellipses are used in many areas of the sciences such as astronomy, physics, engineering, etc. In the 17th century, German mathematician and astronomer Johannes Kepler discovered that all planets orbit around the sun in an elliptical fashion. Below is a figure of an ellipse.

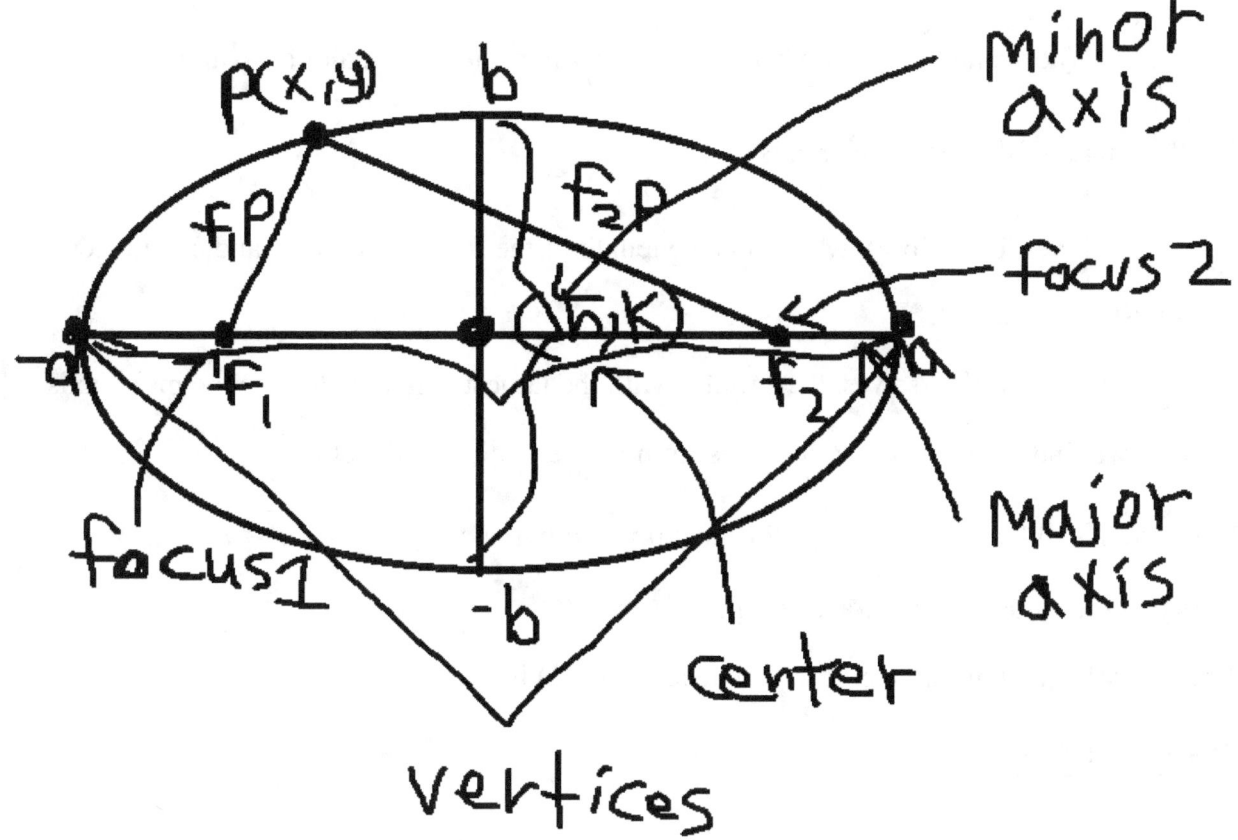

Now the derivation of the ellipse equation is rather long and tedious, which can be left as an exercise for the student. But we can give a quick sketch of what the derivation would look like, which can help us in the matter. In order to start such a calculation of the equation of an ellipse, we need to use the definition of the ellipse. That is, we need to add up the distances from both foci to any point on the ellipse. Referring to the figure above, we can add the distances of the segments.

This leads us to the equation

$f_1p + f_2p = 2a$. Now the length of the major axis is $2a$.

For simplicity, we can let the center of the ellipse be at the origin and let the coordinates of foci f_1 and f_2 be $(-c, 0)$ and $(c, 0)$ respectively.

Applying the distance formula, we have the equation.

$$\sqrt{(x+c)^2 + y^2} + \sqrt{(x-c)^2 + y^2} = 2a$$

After several omitted steps of algebra, we arrive to our equation of the ellipse in standard form.

$\frac{x^2}{a^2} + \frac{y^2}{b^2} = 1$ (horizontal ellipse), where a is the length of the semi-major axis, and

b is the length of the semi-minor axis.

$\frac{x^2}{b^2} + \frac{y^2}{a^2} = 1$ (vertical ellipse), where a is the length of the semi-major axis, and b is the length of the semi-minor axis.

To make it easier for ourselves, the variable with the largest denominator has the major axis.

How do we find the value of c? It comes from the derivation of the ellipse.

When we do all the algebraic steps for the ellipse equation, we get $c = \sqrt{a^2 - b^2}$.

Note that the value of c is always non-negative.

The general equation for an ellipse with a center (h, k) is

$$\frac{(x-h)^2}{a^2} + \frac{(y-k)^2}{b^2} = 1.$$

Another important feature of the ellipse is its eccentricity, which is defined as the ratio of the distance from the center of an ellipse to any focus to the length of the semi-major axis. The eccentricity tells us how normal or unusual the ellipse physically appears. Does it look circular, oval, elongated, or close to a straight line?

The equation for eccentricity of an ellipse is $e = c / a$, where $0 < e < 1$.

THE HYPERBOLA

If we take two cones where one on the bottom is facing up and the one on top is facing down (like the shape of an hour glass), then we can slice both cones vertically to produce a pair of curves. These pair of curves is known as a "hyperbola." The hyperbola is mathematically defined as the set of all points in the plane where the difference between the distances of two fixed points to any point on the hyperbola is a constant. The hyperbola shares important features like the ellipse such as vertices, foci, and eccentricity. On the other hand, a hyperbola is not a closed and bounded curve that we saw earlier in the ellipse. The hyperbola is composed of a pair of curves known as branches. The points where the branches open horizontally or vertically are known as the vertices. Unlike the foci in the ellipse, which were located within the vertices, the foci of the hyperbola are located beyond the vertices. The axis where the vertices, foci, and branches open out is called the transverse axis. The midpoint between the vertices on the transverse axis is the center. Hyperbolas are used in many areas of advanced geometry and physics, especially when studying the motion of a spacecraft and comets relative to the sun or moon. See the figure of a hyperbola below.

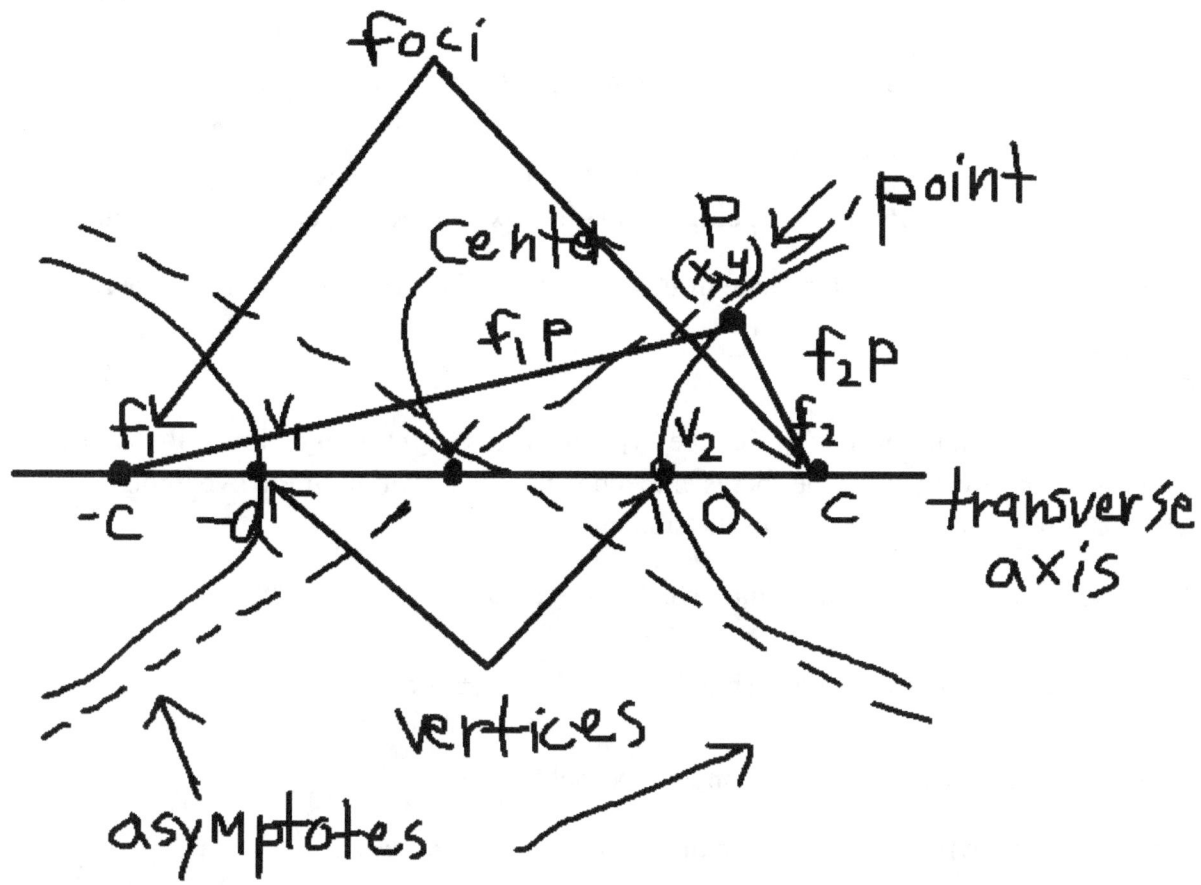

The hyperbola has another important feature. Notice the dotted, imaginary lines that are slanted in the figure above. These imaginary lines are known as asymptotes. This information tells us how the branches behave as the value of x approaches a very large or very small number. We will see that the equations of these asymptotes represent a line in slope-intercept form.

How do we derive the equation for a hyperbola? It is very similar to the derivation of the ellipse equation. We start with taking the difference of the distances f_1p and f_2p. Again, we assume the center to be at the origin for simplicity.

In doing this, we have

$f_1p - f_2p = 2a.$

Applying the distance formula, this gives us

71

$$\sqrt{(x+c)^2 + y^2} + \sqrt{(x-c)^2 + y^2} = 2a.$$

Similar to the ellipse, we take several steps of algebra to get the equation for the hyperbola.

At the end of the day, we have our equation.

$\dfrac{x^2}{a^2} - \dfrac{y^2}{b^2} = 1$, where a is the length of the semi-major/transverse axis, and
b = length of the semi-minor axis.

This equation is very similar to the ellipse equation, except the operation is subtraction.

How do we find the foci or the value of c? The formula is very similar to the ellipse.

$$c = \sqrt{a^2 + b^2}$$

This equation reminds of how to find the hypotenuse of a right triangle in the Pythagorean Theorem! The equation for the foci is embedded in the derivation of the hyperbola formula, which can be left as an exercise for the student.

The eccentricity is the same formula, which is $e = c / a$.

The asymptotes can be found by using the formula $y = \pm \dfrac{b}{a} x$.

These equations are true for a horizontal hyperbola.

For the vertical hyperbola, we switch the x and y variables to get

$$\dfrac{y^2}{a^2} - \dfrac{x^2}{b^2} = 1.$$

And the formula for the asymptotes is $y = \pm \dfrac{a}{b} x$.

In general, for any hyperbola with center (h, k) we have the horizontal and vertical equations for the hyperbola.

$\dfrac{(x-h)^2}{a^2} - \dfrac{(y-k)^2}{b^2} = 1$ (horizontal hyperbola)

$\dfrac{(y-k)^2}{a^2} - \dfrac{(x-h)^2}{b^2} = 1$ (vertical hyperbola)

A special type of hyperbola known as the "standard hyperbola" has the equation $y = c / x$, where c is any non-zero real number.

This concludes the review on conic sections.

COMPUTATIONAL EXERCISES #6

1. Find the focus, vertex, axis of symmetry, and directrix for each of the following parabolas. Graph each parabola.

   a) $x^2 = 12y$

   b) $y^2 = 8x$

   c) $y = 2x^2 - 4x$

   d) $3x = -y^2 + 2y - 10$

2. Find the center and radius of each circle. Graph the circles.

   a) $x^2 + y^2 = 9$

   b) $(x - 2)^2 + (y + 3)^2 = 16$

   c) $x^2 + 10x - 4y + y^2 + 25 = 32$

   d) The endpoints of a circle are (-8, 6) and (2, -2).

3. Find the center, vertices, foci, and eccentricity of the following ellipses. Graph each ellipse.

   a) $\dfrac{x^2}{4} + \dfrac{y^2}{36} = 1$

   b) $\dfrac{(x-1)^2}{9} + \dfrac{(y+4)^2}{25} = 1$

   c) $4x^2 + 25y^2 = 100$

   d) $7x^2 + 8y^2 - 14x + 32y = 17$

4. Find the center, vertices, foci, and asymptotes of each hyperbola. Graph the hyperbola.

a) $\dfrac{x^2}{64} - \dfrac{y^2}{49} = 1$

b) $y^2 - \dfrac{x^2}{25} = 1$

c) $24x^2 = 12y^2 - 24y - 48$

e) The length of the transverse axis for a horizontal hyperbola centered at the origin is 4 units and $c = 3$.

5. Find an equation of a parabola with a focus of $(1, 2)$ and the directrix is $y = -4$.

6. Find an equation of a circle whose diameter is 30 units and center is $(-5, 7)$.

7. Find an equation of an ellipse where the coordinates of the foci are $(-3, 6)$, $(-3, 2)$ and the length of the major axis is 10 units.

8. Find an equation of a hyperbola where the coordinates of the vertices are $(8, 4)$, $(-4, 4)$ and the length of the minor axis is 6 units.

9. What is the length of the latus rectum for the parabola $y = 5x^2 + 20x - 15$?

CONCEPTUAL EXERCISES #6

1. True or False: The distance from the parabola's vertex to the directrix is the same distance from the vertex to its focus.

2. True or False: The center of a circle can be found by completing the square with both x and y variables in any circle equation.

3. True or False: An ellipse where the y-axis is the major axis is called a horizontal ellipse.

4. True or False: A hyperbola has two branches, where one branch is located on the transverse axis, and the other branch is located on the minor axis.

5. True or False: The eccentricity of an ellipse can be found by the equation $e = c / a$.

6. True or False: A circle has no focus.

7. True or False: The latus rectum is the line segment that passes through the focus and has a length of $2p$.

8. True or False: The equation $y = 1 / x$ is an example of a standard hyperbola.

9. Using a diagram and the definition of a parabola, show that the equation for a horizontal parabola centered at the origin is $y^2 = 4px$.

10. Verify that the equation of a circle with radius r centered at the origin is $x^2 + y^2 = r^2$.

11. Derive the ellipse equation. $\dfrac{x^2}{a^2} + \dfrac{y^2}{b^2} = 1$

12. Derive the hyperbola equation. $\dfrac{x^2}{a^2} - \dfrac{y^2}{b^2} = 1$

13. If L is the length of the latus rectum for any parabola, then show that $L = 4p$.

14. Prove that the eccentricity of a hyperbola is $e = \sqrt{1 + \left(\dfrac{b}{a}\right)^2}$.

15. Show that the equation of the asymptotes for a horizontal hyperbola is $y = \pm \dfrac{b}{a} x$ and the equation of the asymptotes for a vertical hyperbola is $y = \pm \dfrac{a}{b} x$.

REVIEW #7

STATISTICS

In this review, we are going to look at the basic statistical measurements such as mean, median, mode, range, and standard deviation to help us analyze data sets. Statistics is commonly used in many areas such as engineering, physics, and other sciences. In addition to reviewing these concepts, we are going to examine other important topics such as the counting principle, permutations, combinations, and probability. Then we will wrap up the review with some logical foundations on statements, truth tables, and set theory. But first, let us begin with the basic statistical measurements.

MEAN: The mean of a data set is like taking the average of all of the numbers in that data set. In other words, we sum up all of the numbers and then divide by the size of the data set.

MEDIAN: The median of a data set is the "middle number" that exists in that data set. Depending on whether we have an odd or even size data set, we can find the median in a couple of ways.

SIZE OF DATA SET IS ODD: If the size of our data set is odd, we can arrange the numbers from least to greatest and cross out the first and last numbers in the data set. We repeat this process until there is only one number left. The number that survives is the median.

SIZE OF DATA SET IS EVEN: Analogous to the odd size data set, we arrange the numbers from least to greatest and cross out the first and last numbers in the data set. However, we will have two numbers remaining instead of one. So, we take the mean or average of those two numbers. This number is the median.

MODE: The mode is the number(s) that occur most frequently in any data set. Some data sets may have only one mode while others may have two modes, three modes, or even no mode at all. If a data set has two modes, then we call it "bimodal." If a data set has three modes, then we call it "tri-modal" and so forth.

RANGE: The range of a data set is the difference between the highest number and the lowest number.

STANDARD DEVIATION: The standard deviation of a data tells us on average how far apart each of the numbers is from the mean. Below is the basic formula for computing the standard deviation of a data set.

$$s = \sqrt{\frac{\sum_{i=1}^{N}(x_i - x^*)^2}{N-1}}$$, where N is the size of the data set, x_i is the ith number in the data set, and x^* is the mean of the data set.

COUNTING PRINCIPLE, PERMUTATIONS, AND COMBINATIONS

Most people have a working knowledge of how to count objects with numbers. But when there are multiple things to count in very large quantities, we need a more efficient way of making an accurate and precise count. We may want to know how many possible ways we can wear a suit given a certain number of shirts, ties, pants, and shoes. Or we may want to know the number of ways to create a password given the number of digits and letters that can be used in a certain manner. There are many applications where we can use these methods of counting. Let us review the foundation of counting, which starts with the counting principle.

THE COUNTING PRINCIPLE

When we count objects, we have a tendency to add up all of the number of objects together to give us the total. The counting principle takes this a step further and uses multiplication instead of addition. We can view multiplication as repeated addition, which helps us to count at a faster rate.

Suppose that we have m ways of doing one thing and n ways of doing another thing. The counting principle says that the total number of ways to do both m and n together is $m*n$. All we did was multiply m and n together to get the total number of ways to do both.

Now the counting principle can be extended to more than just doing two actions. We can do three, five, ten or many actions at a time and determine the number of ways to do perform all of the actions together. All we do is identify the number of ways to do each action and multiply them all together to get the total number of ways.

PERMUTATIONS

Sometimes we want to know how many ways to arrange objects in a certain order. This is where the concept of permutations arises. A permutation is the act of permuting or rearranging a number of objects in a particular way. One important fact that we need to know when dealing with permutations is that the order in which the objects are arranged matters! Here are a couple of formulas that we can use when dealing with permutations.

PERMUTATIONS WITH n OBJECTS: Let n be the number of objects and P_n be the number of ways to arrange n objects. Then $P_n = n!$

PERMUTATIONS WITH n OBJECTS TAKEN k AT A TIME: Let n be the number of objects and k be the number of objects selected to be arranged, where $k \leq n$. Then the number of ways to arrange n objects taken k at a time is $_nP_k = \dfrac{n!}{(n-k)!}$. The term $_nP_k$ can be read as "n permute k."

COMBINATIONS

Sometimes we want to know the number of ways to choose a certain number of objects less than or equal to the total number of objects in a set. We should all be familiar with this formula below since it was the formula we used to establish the binomial coefficients in the binomial theorem.

COMBINATION FORMULA: Let n be the number of objects and k be the number of objects chosen at a time, where $k \leq n$. Then the number of ways to choose k objects at a time out of n objects is $_nC_k = \dfrac{n!}{k!(n-k)!}$. The term $_nC_k$ can be read as "n choose k."

PROBABILITY

There are situations when we want to know the likelihood of an event can happen. What are the chances you can land 2 heads and 1 tail for three coin tosses? What are the odds for you winning the lottery or getting struck by lightning? (Chances are very slim, but it can happen when you least expect it). And so forth. The study of probability will help us to answer these types of questions. First of all, we need to define what probability is.

Definition of probability: Let E be any event that can happen, S be the number of successful outcomes, and let N be the total number of possible outcomes. Then the probability of an event occurring $P(E) = S / N$.

All we do to find the probability of an event occurring is take the number of successful outcomes and divide it by the total number of possible outcomes.

PROBABILITY SUM PRINCIPLE: The sum of the probability of an event happening and the probability of an event not happen is always equal to 1.

That is $P(E) + P(\text{NOT } E) = 1$.

Now there are other formulas we need to know for probability for two independent events. Independent events are where the probability of one event does not affect the probability of another event; e.g. rolling a die and tossing a coin. On the other hand, dependent events are where the probability of subsequent events "depend" on the probability of previous event; e.g. choosing multiple cards without replacing them back in the deck.

Below are the addition and multiplication rules for probabilities for two events.

Let A and B be two distinct, independent events.

ADDITION RULE: $P(A \cup B) = P(A) + P(B)$ The term on the left hand side of the equation is read as "probability of A or B." The probability of events A or B occurring is the sum of their individual probabilities.

Note: Events A and B are "mutually exclusive." This means that A and B do not happen at the same time.

MULTIPLICATION RULE: $P(A \cap B) = P(A)P(B)$ The term on the left hand side of the equation is read as "probability of A and B." The probability of events A and B occurring is the product of their individual probabilities.

NON-MUTUALLY EXCLUSIVE RULE: If A and B are non-mutually exclusive or can happen at the same time, then the addition rule is modified to the following formula below.

$$P(A \cup B) = P(A) + P(B) - P(A \cap B)$$

The probability of non-mutually exclusive events A and B occurring is the sum of their individual probabilities minus the probability both of A and B occurring as stated in the multiplication rule.

The addition, multiplication and non-mutually exclusive rules can be extended to more than two independent events. But these formulas are outside the scope of this text.

STATEMENTS AND TRUTH TABLES

In mathematics, we like to communicate ideas formally with statements. A statement is a sentence that is either true or false, but not both. Some written ideas that are not statements are opinions, interrogatives, and exclamations. Statements either represent truth or falsity. There are logical connectives or operators where we can connect two or more statements together using the OR/AND conjunctions. The OR operator is symbolically expressed as "v," and the AND is symbolically expressed as "^." Statements can also be negated, which yield the opposite truth value. The negation operator is represented by "~."

Like equations that represent an English statement in mathematics, how can we symbolically write a statement? Consider the following below.

Let P and Q be distinct, arbitrary statements. View the table of the English and symbolic representations of statements.

ENGLISH	SYMBOLIC
P or Q	P v Q
P and Q	P ^ Q
not P	~P

Another question we may want to ask is how can we tell whether a given statement is true or false? The answer is by constructing truth tables.

For example, let us view the truth tables for the OR/AND statements and negation of a statement. Note that the final result in each truth table is highlighted in green.

P OR Q: Notice in this truth table below that P or Q is true if both P and Q are true or either P or Q is true. Otherwise, P or Q is false.

P	Q	P v Q
T	T	T
T	F	T
F	T	T
F	F	F

P AND Q: Notice in this truth below that P and Q is true only if both P and Q are true. Otherwise, P and Q is false.

P	Q	P ^ Q
T	T	T
T	F	F
F	T	F
F	F	F

NOT P: Notice in this truth table below that P and NOT P have opposite truth values.

P	~P
T	F
F	T

This leads us to another question. What is the negation of OR/AND statements?

The negation of an OR statement becomes an AND statement with both statements P and Q being negated.

That is ~(P v Q) ≡ ~P ^ ~Q. The symbol ≡ means "logically equivalent to."

And the negation of an AND statement becomes an OR statement with both statements P and Q being negated.

That is ~(P ^ Q) ≡ ~P v ~Q.

A statement where one statement is logically equivalent to another statement like any of the above is called a "tautology." A tautology is always true for every case.

On the other hand, a statement such as P ^ ~P is a statement that is false for every case. Such a statement is known as a "contradiction."

There are other types of statements that we are interested in besides the AND/OR and NOT statements. They are known as the "IF THEN" statements or sometimes called conditional statements.

P → Q means "If P then Q" or "P implies Q," where P is the antecedent and Q is the consequent. The symbol → represents an implication, which indicates that one statement leads to another a final statement or conclusion.

IF P THEN Q: The conditional statement is true for every case except when P is true and Q is false.

P	Q	P → Q
T	T	T
T	F	F
F	T	T
F	F	T

Note: If Q then P is known as the converse of the statement If P then Q. All we do to obtain the converse is switch the antecedent and the consequent.

Another conditional statement known as the bi-conditional statement uses the following tautology.

$(P \leftrightarrow Q) \equiv (P \to Q) \wedge (Q \to P)$

In English, this says, "P if and only if Q is logically equivalent to if P then Q and if Q then P."

The bi-conditional statement breaks up the statement as an AND statement with the implication and its converse.

We will leave it up to the student to prove this is a tautology as an exercise.

How about the negation of an implication? Some of us many think the negation of an implication is another implication. In actuality, it is not. Here is the tautology below.

$\sim(P \to Q) \equiv P \wedge \sim Q.$

To negate the implication, we turn the implication into an AND statement and negate only the consequent. The student can show this is true as an exercise.

Now that we have covered all of the basic statements and their respective truth tables, there is one more thing we need to cover. And that is how to relate the number of statements to the number of cases we need to construct in our truth table. When we had only one statement P, we had 2 cases. On the other hand, when we had two statements P and Q, we had 4 cases or 2^2. If we had three statements P, Q, and R, we would have 8 cases or 2^3. So, it appears that the number of cases needed for n number of statement is 2^n. Imagine constructing a truth table for $n \geq 5$. Such a construction would be very tedious and time consuming!

SET THEORY

In the real world, we all know what a set of objects looks like. The game of pool has a set of solid balls and striped balls. A classroom contains a set of students and a teacher. The mathematical definition of a set is a collection of elements. Some sets contain finite number of elements while other sets contain an infinite number of elements.

If we want to say that a certain element x belongs to a set S, we can denote this with the following notation.

$x \in S$: "x is in the set S."

Some elements may not belong in a set. Thus, we can denote this as

$y \notin S$: "y is not in the set S."

Now the set that contains no elements is called the "empty set" or symbolically write as $\varnothing$.

A set may have some or all of their elements that also belong to another set. That is, if every element in a set A also belongs to another set B, then we say "A is a subset of B." Here is the mathematical definition below.

$A \subseteq B$: If $x \in A$ then $x \in B$.

If B is a subset of A, then we switch A and B.

$B \subseteq A$: If $x \in B$ then $x \in A$.

Now what if A is a subset of B and B is a subset of A? Then we know that the two sets are equal. So, $A = B$.

It is important to know that the empty set is a subset of any other set.

We are going to introduce some basic set operators that every student should know.

SET OPERATORS

UNION OF SETS: Let A and B be two distinct sets. The union of A and B is defined below.

$A \cup B$: $x \in A$ or $x \in B$

Notice that the union uses the OR operator.

INTERSECTION OF SETS: Let A and B be two distinct sets. The intersection of A and B is defined below.

$A \cap B$: $x \in A$ and $x \in B$

Notice that the intersection used the AND operator.

DIFFERENCE OF SETS: Let A and B be two distinct sets. The difference of A and B is defined below.

$A \setminus B$: $x \in A$ and $x \notin B$

COMPLEMENT OF SETS: Let U be the universal set and let $A \subseteq U$. The complement of A is defined below.

A^C: $x \in U$ and $x \notin A$.

Note that the similarities between the difference of two sets and the complement of one set. The only difference is that complement usually deals with the universal set, which is frequently a larger set than sets A and B.

Below are some important set properties that can be useful when proving statements with sets.

SET PROPERTIES:

COMMUTATIVE PROPERTY: $A \cup B = B \cup A$, $A \cap B = B \cap A$

ASSOCIATIVE PROPERTY: $A \cup (B \cup C) = (A \cup B) \cup C$, $A \cap (B \cap C) = (A \cap B) \cap C$

DISTRIBUTIVE PROPERTY: $A \cup (B \cap C) = (A \cup B) \cap (A \cup C)$,
$A \cap (B \cup C) = (A \cap B) \cup (A \cap C)$

IDENTITY PROPERTY: $A \cup \varnothing = A$

SYMMETRIC DIFFERENCE PROPERTY: $A^\Delta B = (A \cup B) \setminus (A \cap B)$

This review section is definitely longer in relation to the previous review sections in the text. And the material presented here is more abstract. So, feel free to read over it multiple times to help you grasp the concepts before you move on to the exercises.

COMPUTATIONAL EXERCISES #7

1. Find the mean, median, mode, range, and standard deviation of the data set {3, 12, 8, 7, 4, 1, 3}.

2. John has test scores of 85, 60, 72, and 90 in his math class. What minimum score does John need to get to have an average of at least an 80?

3. How many different sandwiches can you make with 3 different types of bread, 5 different cheeses, and 4 kinds of meat?

4. How many ways can you arrange 5 books on a bookshelf?

5. Out of a set of 10 marbles, how many ways can you choose 7 of them?

6. A password is composed of 3 digits and 3 letters in that order. Assuming that the digits and letters cannot be repeated, how many possible passwords can be created?

7. When rolling two dice, what is the probability of rolling a 5 or a 12?

8. What is the probability of drawing a jack or a spade in a deck of 52 cards?

9. Using a spinner labeled 1-4 and an un-weighted coin, what is the probability of landing tails and spinning an odd number?

10. In a local lottery, you can choose six numbers from 1-30. The order in which you choose the numbers does not matter. What are your chances of winning the lottery?

11. In a game, a contestant is blindfolded so he or she cannot see the marbles to choose from. There are 2 green marbles, 3 blue marbles, and 4 red marbles. The contestant chooses 2 marbles without replacement. If the contestant chooses 2 marbles that are different colors, then he or she wins the game. What is the probability of the contestant winning the game? Losing the game?

12. Construct truth tables for $P \wedge \sim Q$, $\sim P \rightarrow Q$, and $P \vee (Q \wedge R)$.

13. Construct truth tables for $P \rightarrow (Q \vee R)$ and $(P \wedge \sim R) \rightarrow Q$. What do you notice?

14. You have statements P, Q, R, S, T, and V. How many possible cases do you need to present in your truth table?

15. Determine whether each of the following statements is true or false. If the statement is true, prove it. If the statement is false, give a counterexample to show why it is false.

a) All prime numbers are odd integers.

b) If n is an even number, then $n^2 + 2n$ is even.

c) $x = 5$ if and only if $x^2 = 25$

d) The sum of three consecutive integers is always divisible by 3.

e) If t is an odd number, then t^2 is odd.

16. Let $A = \{1, 2, 3\}$, $B = \{2, 3, 4\}$, $C = \{5, 6\}$, and $U = \{1, 2, 3, 4, 5, 6\}$. Find each of the following sets.

a) $A \cup B$

b) $B \cap C$

c) $B \setminus A$

d) A^C

e) $B \cup (A \cap C)$

f) $(A \cup B)^C \setminus C$

CONCEPTUAL EXERCISES #7

1. True or False: The way to find the mean of a data set is to find the middle number in that data set.

2. True or False: The range is the difference between the highest and lowest values in a data set.

3. True or False: Permutations are used in situations where the order of objects matters.

4. True or False: Combinations are used in situations regardless of whether the order of objects chosen matters.

5. True or False: "There are over 10,000 lakes in Minnesota" is an example of a statement.

6. True or False: The negation of an OR statement is an AND statement. And both separate statements are negated.

7. True or False: Both the IF THEN statement and its converse are always true.

8. True or False: $\sim P \vee Q \equiv P \rightarrow Q$

9. True or False: The set $\{\varnothing\}$ is considered to be an empty set.

10. True or False: In order to show that two sets are equal, you need to show that one set is a subset of the other set and vice versa.

11. True or False: The intersection of two disjoint sets that have no common elements is considered to be an empty set.

12. True or False: If $A = \{a, b, c, d\}$, $B = \{b, d, e\}$, and $U = \{a, b, c, d, e\}$, then $A \cup (B^C \setminus A) = A$.

13. Let n be the number of people at a dinner party and let H be the number of possible handshakes taken with only 2 people at a time. Show that the number of possible handshakes is $H = n(n-1)/2$.

14. Prove that $P \rightarrow Q \equiv \sim Q \rightarrow \sim P$. This is known as the contrapositive of an implication.

15. Use the contrapositive from exercise 14 to prove the following statement.

    "If p^2 is even, then p is even."

16. The EXCLUSIVE OR STATEMENT $P \oplus Q$ is true when P and Q have opposite truth values and false when P and Q share the same truth values. Determine if the statement $P \oplus (Q \vee R) \equiv (P \oplus Q) \wedge (P \oplus R)$ is a valid equality. If the equality is false, find another equality to make the statement true.

17. If $A = B$, then show that $A \setminus B = \varnothing$.

18. Show that $(A \cap B)^C = A^C \cup B^C$.

19. Let N represent the integers from 1-10, E represent the even integers from 1-10 and F represent the integers from 1-10 that are divisible by 4.

    Prove or disprove the statement $E \triangle F = E \setminus F$.

20. Let $S = \{a, b, c\}$ and $P(S)$ be the power set of S. The power set is the set that contains all possible subsets of the given set. Find all of the elements in $P(S)$.

REVIEW #8

TRIGONOMETRY PART 1

One of the most applied branches in all of mathematics is the study of trigonometry. Trigonometry is the study of triangles and the relationships between the sides and the angles formed between those sides. In this first part of the review, we will cover the main topics learned in trigonometry such as the six trigonometric functions (inverse functions) and their graphs, and the unit circle.

ANGLE MEASUREMENTS AND CONVERSIONS

Angles can be measured in two different ways: 1. Degrees and 2. Radians. The latter is most common when dealing with angles on the unit circle, which will be covered in near future. The question is how do we convert from degrees to radians and vice versa? The answer is simple. See below.

DEGREES TO RADIANS: Multiply the number of degrees by $\pi/180$.

RADIANS TO DEGREES: Multiply the number of radians by $180/\pi$.

It is important to note that 1 radian ≈ 57.3 degrees. If we multiply 57.3 degrees by 3.14, then we have approximately 180 degrees or π radians.

A circle is 360 degrees or 2π radians.

THE SIX TRIGONOMETRIC FUNCTIONS

Any student who has learned trigonometry in high school or college should be familiar with the six trigonometric functions that are used for all right triangles. A trigonometric function reveals the ratio of two sides of a right triangle for a given angle. These functions can help us to find an unknown sides and angles of right triangles. Let us review the six trigonometric functions.

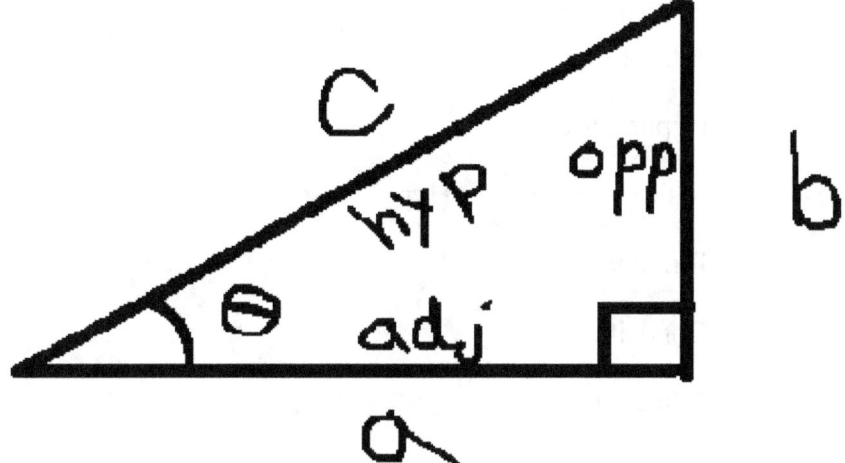

We can see above the right triangle with the angle θ formed by the adjacent side and the hypotenuse. The opposite side is directly across angle θ and the hypotenuse is directly across the 90 degree angle.

SINE: $\sin\theta = b/c$

COSINE: $\cos\theta = a/c$

TANGENT: $\tan\theta = b/a = \sin\theta/\cos\theta$

SECANT: $\sec\theta = c/a = 1/\cos\theta$

COSECANT: $\csc\theta = c/b = 1/\sin\theta$

COTANGENT: $\cot\theta = a/b = 1/\tan\theta$

The first three trigonometric functions are known as the basic functions. The last three trigonometric functions are known as the reciprocal functions.

A great pneumonic device that can help us relate each trigonometric function to their corresponding ratios is the Chief Indian of the right triangles named SOHCAHTOA.

S = Sine

O = Opposite

H = Hypotenuse

So, Sine = Opposite / Hypotenuse

C = Cosine

A = Adjacent

H = Hypotenuse

So, Cosine = Adjacent / Hypotenuse

T = Tangent

O = Opposite

A = Adjacent

So, Tangent = Opposite / Adjacent

And we know that cosecant relates to sine, secant relates to cosine, and cotangent relates to tangent. So, we can switch the letters of pairs OH, AH, and OA from all three trigonometric functions to get the reciprocal functions. Therefore, we have the following.

Cosecant = Hypotenuse / Opposite

Secant = Hypotenuse / Adjacent

Cotangent = Adjacent / Opposite

Before we analyze the graphs of trigonometric functions, we need to learn about how to evaluate the ratios of these trigonometric functions through utilizing the unit circle.

THE UNIT CIRCLE

Recall from Review #6 that the equation of a circle with radius r centered at the origin is $x^2 + y^2 = r^2$. The unit circle is a circle that is centered at the origin with a radius of 1 unit. Since $r = 1$, our equation becomes $x^2 + y^2 = 1$. How can we determine the x and y coordinates in terms of angle θ? The drawing below can provide us some good insight in doing this.

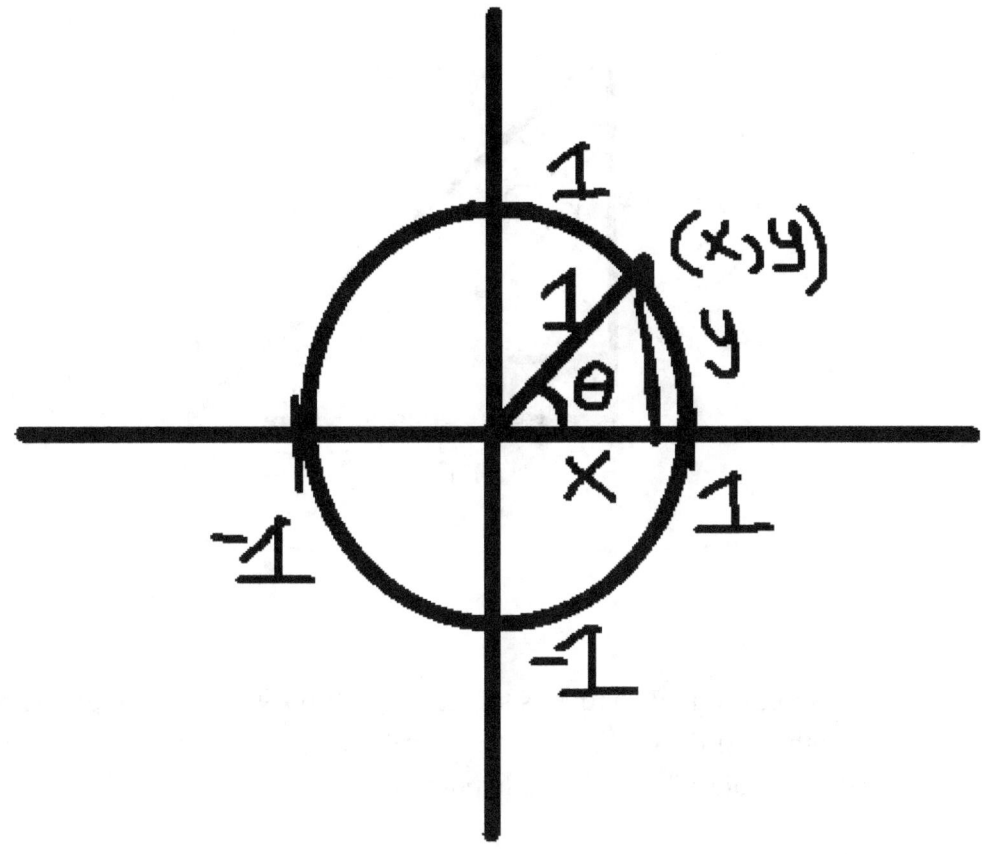

We can draw a right triangle within the unit circle where the angle θ is formed between the adjacent side on the *x*-axis and the hypotenuse with a length of 1 unit.

From our trigonometric functions, we know that cosine = adjacent / hypotenuse and sine = opposite / hypotenuse.

Since *x* = adjacent, *y* = opposite, and hypotenuse = 1, we know that

$x = \cos\theta$ and $y = \sin\theta$.

Since $\tan\theta = \sin\theta / \cos\theta$, we know that

$\tan\theta = y / x$.

Also note that angles which rotate counterclockwise around the unit circle are positive, and angles which rotate clockwise around the unit circle are negative.

Now that we are equipped with this information we can find the sine, cosine, and tangent of some fundamental angles on the unit circle. How do we do this? We refer back to the 30-60-90 and 45-45-90 right triangles.

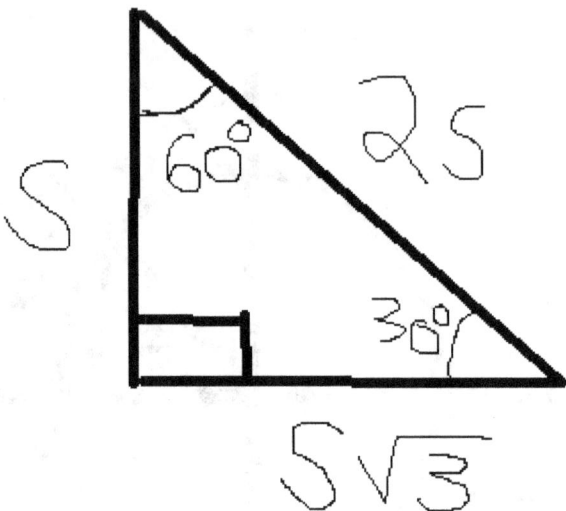

This right triangle above is the general 30-60-90 right triangle with side of length s. Since we have the unit circle, we know that $s = 1$. So, the legs are square root of 3, 1, and the hypotenuse is 2.

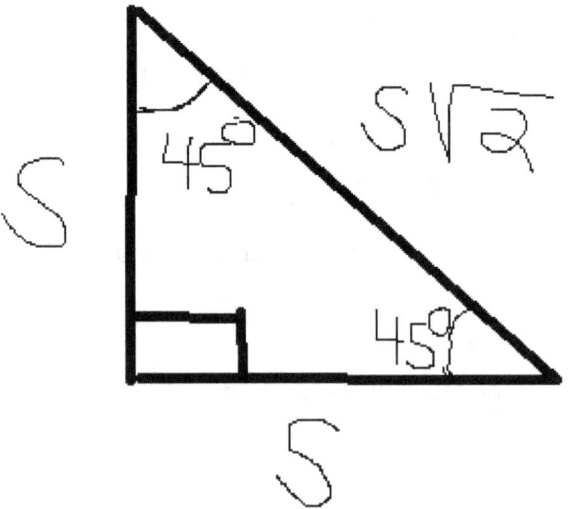

The other right triangle above is the general 45-45-90 right triangle with side of length s. Like before, we let $s = 1$. So, the legs are both 1 and the hypotenuse is the square root of 2.

Gathering all of our data together, we know the following facts.

$$\text{Sin}(30°) = 1/2 \qquad \text{Sin}(45°) = \frac{\sqrt{2}}{2} \qquad \text{Sin}(60°) = \frac{\sqrt{3}}{2}$$

$$\text{Cos}(30°) = \frac{\sqrt{3}}{2} \qquad \text{Cos}(45°) = \frac{\sqrt{2}}{2} \qquad \text{Cos}(60°) = 1/2$$

$$\text{Tan}(30°) = \frac{\sqrt{3}}{3} \qquad \text{Tan}(45°) = 1 \qquad \text{Tan}(60°) = \sqrt{3}$$

The angles 30°, 45°, and 60° that exist in the first quadrant are known as the reference angles. These reference angles will help us to find the other trigonometric ratios of other angles that exist in the remaining three quadrants.

How do we find the sine, cosine, and tangent of other angles like 0°, 90°, 180°, 270°, and 360°? We need to look at the unit circle diagram.

For 0°, there is only a horizontal component or straight angle formed along the positive x-axis. So, cos(0°) = 1 and sin(0°) = 0. This implies that tan(0°) = 0. The same is also true for 360°.

For 90°, there is only a vertical component or a right angle formed by the positive x and y axes. So, cos(90°) = 0, and sin(90°) = 1. This implies that tan(90°) is undefined.

For 180°, there is only a horizontal component or a straight angle formed by the x-axis with the terminal end on the negative x-axis. So, cos(180°) = -1 and sin(180°) = 0. This implies that tan(180°) = 0.

For 270°, there is only a vertical component formed, which ends on the negative y-axis. So, cos(270°) = 0 and sin(270°) = -1. This implies that tan(270°) is undefined.

It is important to know that the angles on the unit circle are defined in terms of radians instead of degrees. The student will be given an opportunity to convert all of the angles on the unit circle to radians as an exercise.

Now that we have a better understanding of the unit circle, we can now view necessary features and properties of the trigonometric graphs.

GRAPHS OF TRIGONOMETRIC FUNCTIONS

It is not only important to understand the ratios for each trigonometric function, but we also need to understand the graphical appearance of trigonometric functions and their behaviors. Trigonometric functions are considered to be periodic functions. This means the functions undergo a period or a cycle of a certain type of behavior, which repeats over and over again. We see periodic (trigonometric) functions in nature everyday such as day and night in a 24 hour period, the low and high tides that occur in the ocean, and the four seasons that we experience every year in life.

We will begin our analysis of trigonometric functions with the sine and cosine graphs.

Below are the graphs of $y = \sin(x)$ in red and $y = \cos(x)$ in green. Notice that the shapes of these graphs represent the form of a wave. The angles on the x-axis are expressed in radians

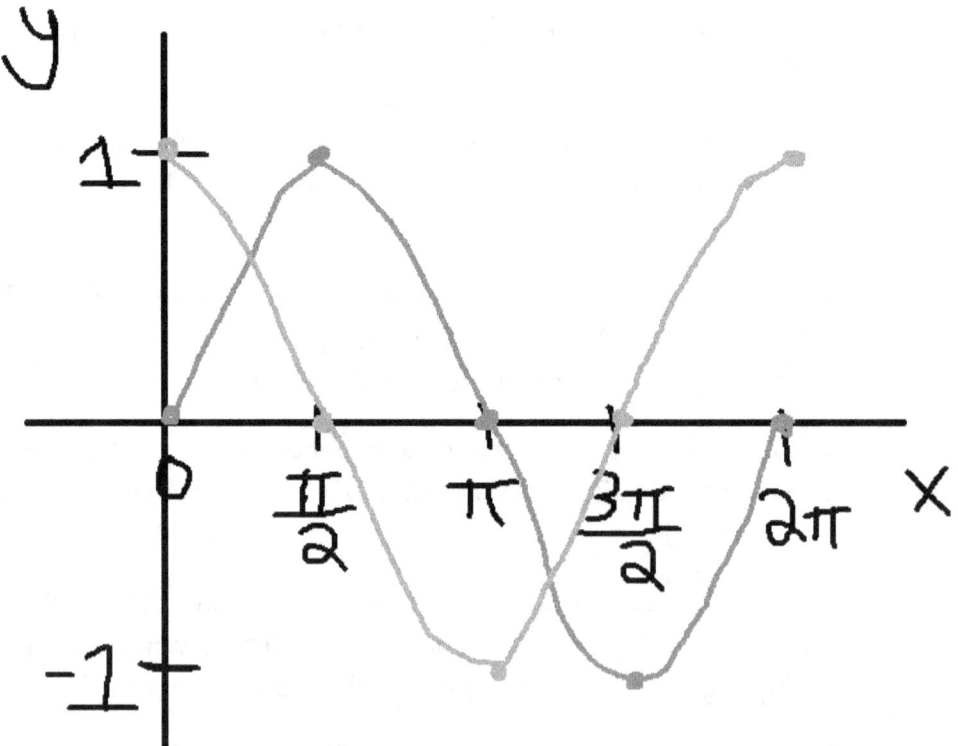

There are three important properties that the sine and cosine waves share.

1. AMPLITUDE: The maximum displacement from the equilibrium position to the crest (high point) or trough (low point) of the wave.

2. PERIOD: The amount of time it takes for a wave to complete one full cycle. Period is measured in seconds.

3. FREQUENCY: The number of cycles performed in one second. Frequency is measured in a unit called Hertz (Hz).

Both the $y = \sin(x)$ and $y = \cos(x)$ have an amplitude of 1, a period of 2π, and a frequency of $1 / 2\pi$.

The domain of sine and cosine are all real numbers. But the range is [-1, 1].

However, the amplitude, period, and frequency can vary based on the given equation. The sine and cosine waves can also shift upwards and downwards.

Here are some general forms of these equations.

FORM 1: $y = A\sin(x)$ and $y = A\cos(x)$.

AMPLITUDE: $|A|$

PERIOD: 2π

FREQUENCY: $1 / 2\pi$

DOMAIN: All real numbers

RANGE: $[-A, A]$

FORM 2: $y = A\sin(Bx)$ and $y = A\cos(Bx)$

All of the attributes are the same as in FORM 1 except the period and frequency.

PERIOD: $2\pi / |B|$

FREQUENCY: $|B| / 2\pi$

FORM 3: $y = A\sin(Bx - C)$ and $y = A\cos(Bx - C)$, where C is the horizontal shift. The phase shift is C / B.

If $C > 0$, then the graph is shifted C-units to the right.

If $C < 0$, then the graph is shifted C-units to the left.

If $C = 0$, then we revert back to FORM 2.

All of the other attributes are the same as in FORM 2.

FORM 4: $y = A\sin(Bx - C) + D$ and $y = A\cos(Bx - C) + D$, where D is the vertical shift.

If $D > 0$, then the graph is shifted D-units upward.

If $D < 0$, then the graph is shifted D-units downward.

If $D = 0$, then we revert back to FORM 3.

The domain is the same as all the other forms, but the range differs.

RANGE: $[D - A, D + A]$

If we take the average of the minimum and maximum values in the range, we see that $(D - A + D + A) / 2 = 2D / 2 = D$. So, D is the equilibrium position where the wave begins. D is also called the "centerline."

These are the general forms for the sine and cosine graphs. The other trigonometric function we want to look at closely is $y = \tan(x)$.

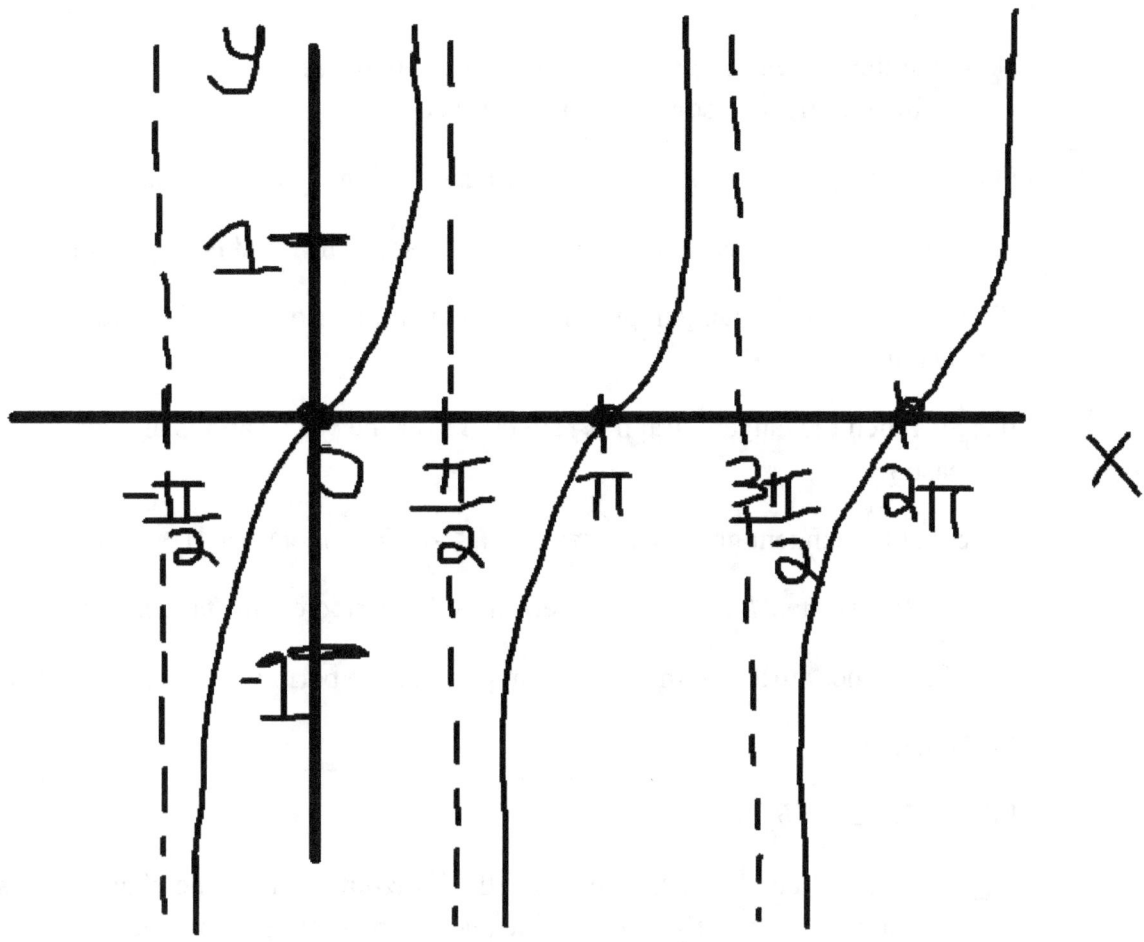

The tangent graph has no amplitude since it does not have a maximum or minimum value like the sine and cosine graphs. However, the tangent graph does have a period, frequency, and may be shifted horizontally and/or vertically as well. We can see that the tangent graph has vertical asymptotes at odd multiples of $\pi / 2$.

Here are the attributes of $y = \tan(x)$ below.

AMPLITUDE: N/A

PERIOD: π

FREQUENCY: $1 / \pi$

DOMAIN: All real numbers except odd multiples of $\pi / 2$

RANGE: All real numbers

Note that the tangent graph can also be written in the last three forms like the sine and cosine graphs with some minor differences.

FORM 1: $y = A\tan(x)$ The value of A is a narrowing or widening factor.

If $A > 1$, then the tangent graph narrows by a factor of A and is always increasing.

If $0 < A < 1$, then the tangent graph widens by a factor of A and is always increasing.

If $A < 0$, then the tangent graph is reflected over the y-axis and is always decreasing.

Note that the tangent graph narrows for $A < -1$ and widens for $-1 < A < 0$.

FORM 2: $y = A\tan(Bx)$ The $|B|$ determines the period of this tangent graph.

All of the other attributes in FORM 1 apply, but the period and frequency differ.

PERIOD: $\pi / |B|$

FREQUENCY: $|B| / \pi$

FORM 3: $y = A\tan(Bx - C)$, where C is the horizontal shift. The phase shift is C / B, which works exactly like the sine and cosine graphs. However, the domain will vary based on the phase shift.

To find the vertical asymptotes, set $Bx - C = (2k + 1)\pi / 2$, where k is any integer. Then solve for x.

So, the domain is all real numbers except $x = (2k + 1)\pi / 2B + C / B$.

FORM 4: $y = A\tan(Bx - C) + D$, where D is the vertical shift. All of the other attributes from the previous two forms apply.

We will omit the other three trigonometric graphs and leave them as an exercise for the student to graph them. Now we will move forward to the inverse trigonometric functions.

INVERSE TRIGONOMETRIC FUNCTIONS

A question we want to ask ourselves about trigonometric functions is do they have an inverse? The answer is yes! But we may wonder why this is so for the sine and cosine graphs since passing the horizontal line test is questionable. Recall that an inverse function reverses the variables x and y, which also reverses the domain and range. We need to restrict values for the domain of the original trigonometric graphs in order to find the range of the inverse trigonometric graphs. Let us examine the inverse sine and inverse cosine graphs first.

INVERSE SINE

When we graph $y = \sin(x)$ in the interval $[0, 2\pi]$, we see that the horizontal line test fails. So, we need to restrict our domain so the horizontal line test passes. If we graph on the interval $[-\pi / 2, \pi / 2]$, we see that the function is one-to-one. So, we can proceed to find the inverse $y = \sin^{-1}(x)$. Here is the graph of the inverse sine function.

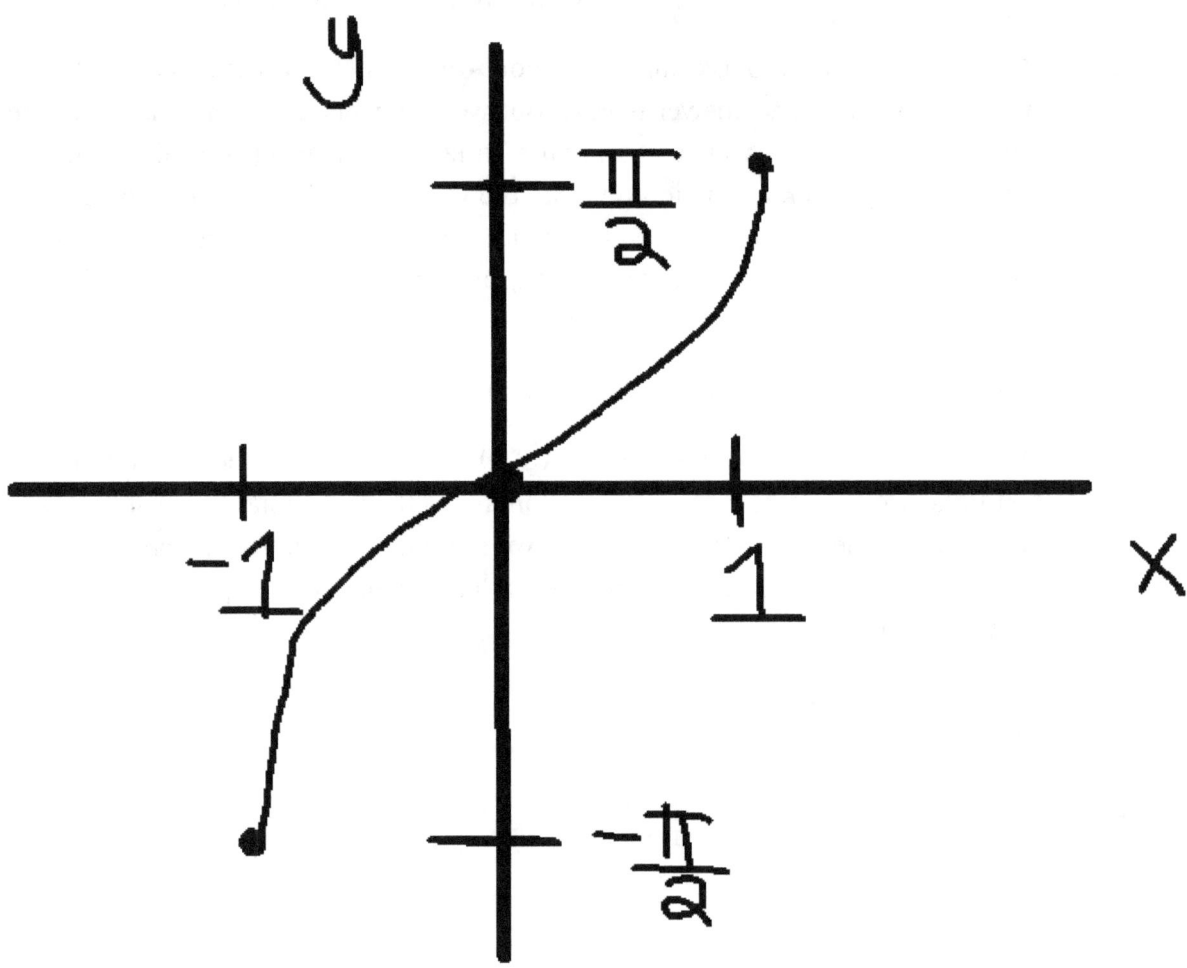

Here are the following attributes of the inverse sine function.

DOMAIN: [-1, 1]

RANGE: [-π / 2, π / 2]

Note that the input value x represents the trigonometric ratio between -1 and 1 inclusively and the output value y represents the angle whose sine of that angle yields that ratio.

Now we move on to the inverse cosine function.

100

When we graph $y = \cos(x)$ on the interval $[0, 2\pi]$, we run into the same problem as before with the inverse sine graph. The graph fails to pass the horizontal line test, so we need to restrict the domain of the cosine function. If we graph the function on the interval $[0, \pi]$, then the graph passes the horizontal line test and guarantees us an inverse. Now we can find the inverse cosine function $y = \cos^{-1}(x)$. Here is the graph.

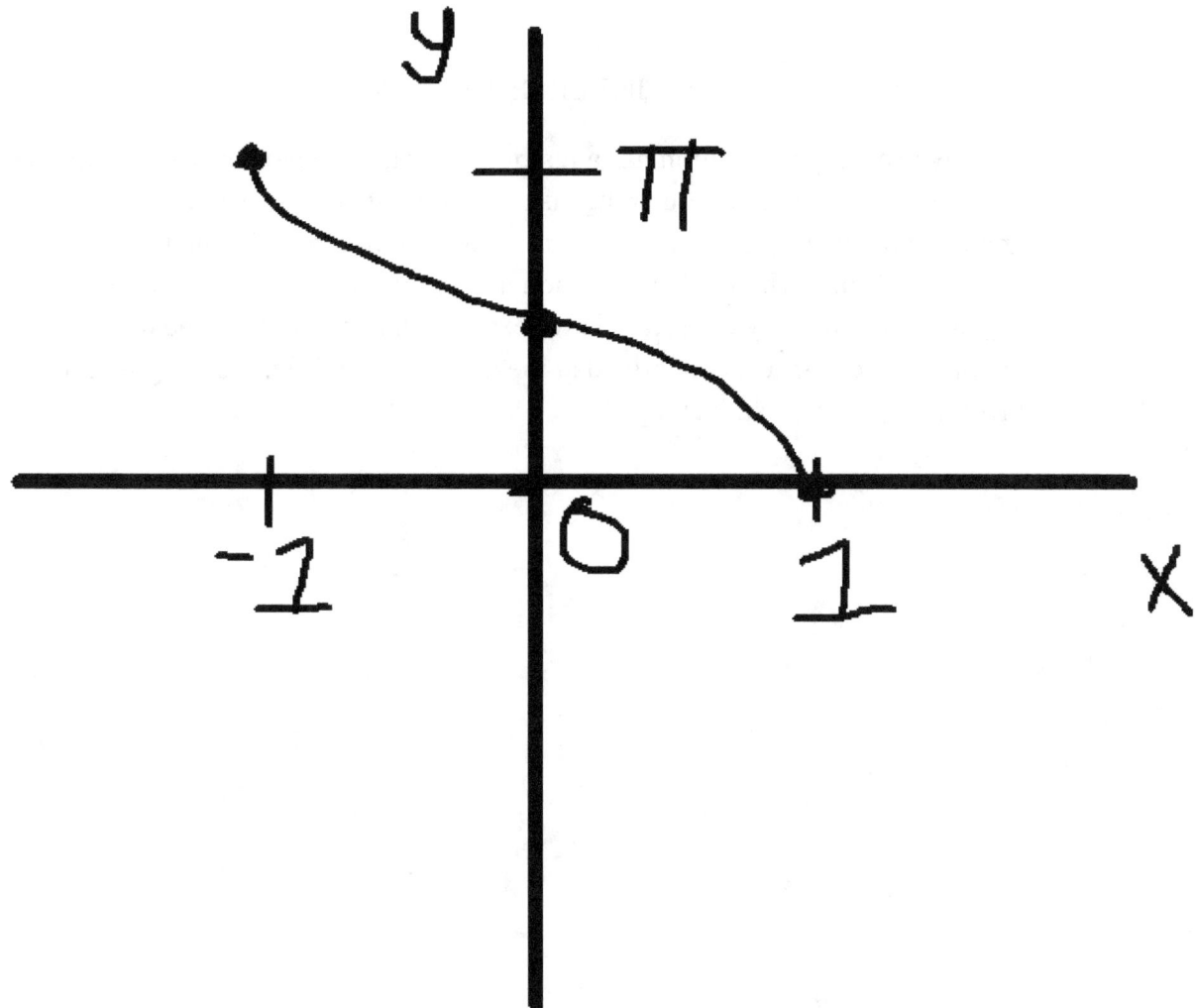

Here are the following attributes of the inverse cosine function.

DOMAIN: [-1, 1]

RANGE: $[0, \pi]$

Notice that the domain of the inverse cosine is the same as the domain of the inverse sine. Of course, their ranges are different. Analogous to the inverse sine function, the input value x represents the trigonometric ratios and the output value y represents the angle whose cosine is that ratio.

Finally, we observe the inverse tangent function.

INVERSE TANGENT

The inverse tangent function has a restricted domain due to vertical asymptotes at odd multiples of $\pi / 2$, but the range is all real numbers. All we have to do to graph the inverse tangent function is restrict the domain to the interval $[-\pi / 2, \pi / 2]$ like with the sine function and graph the curve between the vertical asymptotes. The only exception is we exclude the endpoints of the interval for the range since $\tan(x)$ is undefined at these two angles. Here is the graph of inverse tangent $y = \tan^{-1}(x)$.

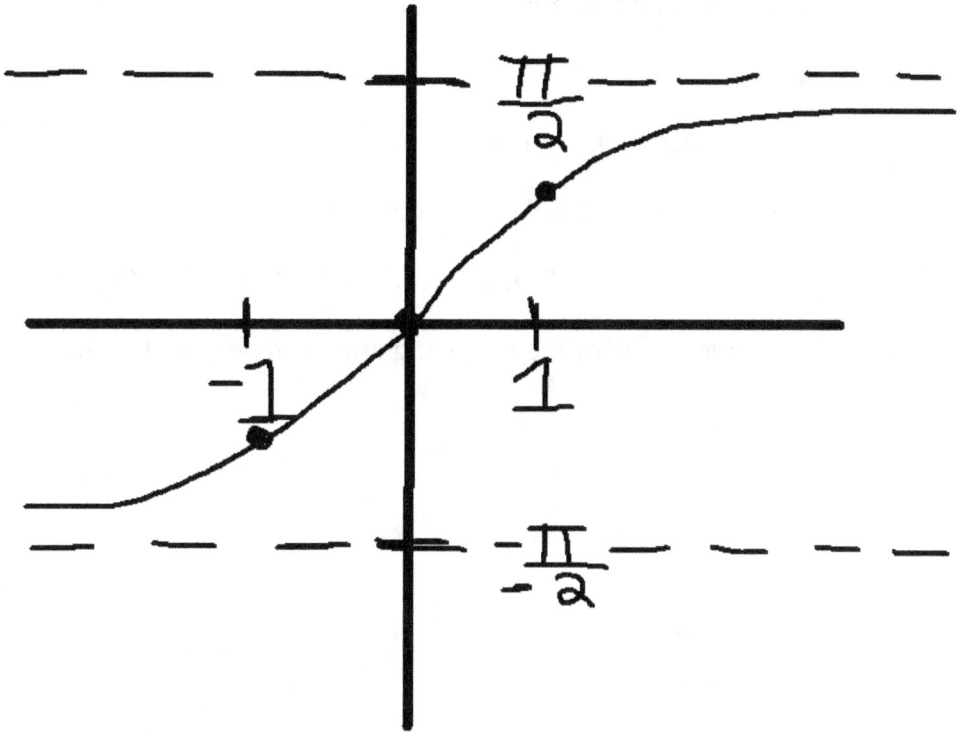

Here are the following attributes of the inverse tangent function.

DOMAIN: All real numbers

RANGE: $(-\pi / 2, \pi / 2)$

The great aspect about the inverse tangent function is we can input any real number for x. But the range is restricted to all values between $-\pi / 2$ and $\pi / 2$. The inverse tangent function tells us to input any real ratio and get the angle whose tangent is that ratio.

For the sake of brevity and practicality we will omit the inverse trigonometric functions for secant, cosecant, and cotangent.

INVERSE TRIGONOMETRIC PROPERTIES

We need to establish a few properties of inverse trigonometric functions before starting to solve trigonometric equations and finding their inverses. Here are these properties.

A. $y = \sin(x)$ if and only if $x = \sin^{-1}(y)$

B. $y = \cos(x)$ if and only if $x = \cos^{-1}(y)$

C. $y = \tan(x)$ if and only if $x = \tan^{-1}(y)$

D. $y = \csc(x)$ if and only if $x = \csc^{-1}(y)$

E. $y = \sec(x)$ if and only if $x = \sec^{-1}(y)$

F. $y = \cot(x)$ if and only if $x = \cot^{-1}(y)$

COMPUTATIONAL EXERCISES #8

1. Convert each of the following angles in degrees to radians.

   a) $20°$

   b) $45°$

   c) $150°$

   d) $-300°$

   e) $-225°$

2. Convert each of the following angles in radians to degrees.

   a) $\pi / 2$

   b) $5\pi / 4$

   c) $-\pi / 3$

   d) $7\pi / 6$

   e) $-2\pi / 3$

3. Evaluate the ratios for all six trigonometric functions for every angle listed on the unit circle and organize the results in a table. Make sure to convert the angle measurements in degrees to radians.

   Angles: $0°$, $30°$, $60°$, $90°$, $120°$, $135°$, $150°$, $180°$, $210°$, $225°$, $240°$, $270°$, $300°$, $315°$, $330°$, $345°$, $360°$.

4. Graph each of the sine graphs in their various forms. Identity the amplitude, period, frequency, phase shift, domain and range for each graph.

   a) $y = 2\sin(x)$

   b) $y = -3\sin(2x)$

c) $y = \sin(x - \pi / 2) + 1$

d) $y = 5\sin(4x + \pi / 3) - 2$

5. Graph each of the cosine graphs in their various forms. Identify the amplitude, period, frequency, phase shift, domain and range for each graph.

   a) $y = -\cos(x)$

   b) $y = 4\cos(6x) + 3$

   c) $y = 2\cos(3\pi x - \pi)$

   d) $y = 8\cos(x) - 5$

6. Graph each of the tangent graphs in their various forms. Identify the period, frequency, domain and range for each graph.

   a) $y = \tan(4x)$

   b) $y = -2\tan(x + \pi / 6)$

   c) $y = 5\tan(x) + 2$

   d) $y = -\tan(8x - \pi / 2) + 1$

7. Find an equation in the form $y = A\cos(Bx - C) + D$ given the following information.

   Maximum = 4, Minimum = -2, Period = $\pi / 4$, Phase Shift = $\pi / 12$

8. Graph each of the reciprocal trigonometric functions. Identify the period, frequency, domain, and range for each graph.

   a) $y = \csc(x)$

   b) $y = \sec(x)$

   c) $y = \cot(x)$

9. Evaluate each of the following for the inverse trigonometric functions on the interval $[0, 2\pi]$.

   a) $\sin^{-1}(1/2)$

b) $\cos^{-1}(0)$

c) $\tan^{-1}(-1)$

d) $\cos(\sin^{-1}(\sqrt{3}/2))$

e) $\cos^{-1}(\tan(-\pi/4))$

10. Find the values of θ where $\sin\theta > 0$, $\cos\theta > 0$, and $\tan\theta > 0$.

11. If $\sin\theta = 3/5$, then what is the value of:

a) $\cos\theta$?

b) $\tan\theta + \sec\theta$?

c) $\csc\theta - \cot\theta$?

CONCEPTUAL EXERCISES #8

1. True or False: For any point (x, y) and given angle on the unit circle, the x coordinate can be found by taking the cosine of the given angle and the y coordinate can be found by taking the sine of the given angle.

2. True or False: The pneumonic SOHCAHTOA can be used to memorize the ratios of all six trigonometric functions.

3. True or False: To convert from radians to degrees, you multiply the angle in radians by $\pi / 180$.

4. True or False: The amplitude of the wave for a trigonometric graph is the minimum displacement of the wave from its equilibrium position.

5. True or False: The period and frequency are inversely related to one another.

6. True or False: The period of $y = \tan(x)$ is the same as the period of $y = \sin(x)$ and $y = \cos(x)$.

7. True or False: The domain of both the inverse sine and inverse cosine functions are equal, but their ranges are different.

8. True or False: $y = \sin(x)$ is an odd function.

9. True or False: $y = \cos(x)$ is an odd function.

10. True or False: $y = \tan(x)$ is an even function.

11. If $y = A\sin(2\pi Ax - A) + A$, then what is the domain, range, amplitude, period, and frequency for this sine function?

12. What is the inverse function of $f(x) = A\cos(Bx)$? Verify that $f^{-1}(A) = 0$.

13. If $\theta = \sin^{-1}(x^{1/2})$, prove that $(x\sec\theta)^2 + x = \tan^2\theta$

14. For what values of k does $\cos(k\pi) = 1$ or -1?

15. For what values of k does $\sin(k\pi) = 0$?

16. Find a general equation for the vertical asymptotes of $y = \tan(Bx + B)$ in terms of π and the period T.

17. The trigonometric function $v(t) = -Aw\sin(wt)$ describes the velocity of a wave, where the angular frequency $w = 2\pi f$ (f is the standard frequency of the wave). Show that the wave achieves maximum height at ¾ of the wave's period. What is the maximum velocity of the wave at that instant?

18. If $\cos(A) = y$, $\sin(B) = x$, and $p = (1 - y^2)^{1/2}$, $q = (1 - x^2)^{1/2}$, then prove that
$$\frac{1}{\sin(A) - \cos(B)} = \frac{1}{p - q}.$$

REVIEW #9

TRIGONOMETRY PART 2

The second part of the trigonometry review will expand upon the concepts we learned earlier in the first part. Part 2 of the review will cover trigonometric identities, Law of Sines, and Law of Cosines. Let us review the equation of the unit circle to derive some important trigonometric identities.

Recall that the unit circle is centered at the origin with a radius of 1 unit. So, the equation of the unit circle is $x^2 + y^2 = 1$. Now we know how to relate x and y to the cosine and sine functions respectively.

So, $x = \cos\theta$ and $y = \sin\theta$. If we substitute these equations into our unit circle equation above, we will have $\cos^2\theta + \sin^2\theta = 1$. This equation is true for all angles on the unit circle. We call this equation the Pythagorean Identity.

Now there are two other Pythagorean Identities we can derive from the first one we found above. If we divide by $\cos^2\theta$ to both sides of this equation, we will get $1 + \tan^2\theta = \sec^2\theta$. This is the second Pythagorean Identity. The third identity is found by dividing by $\sin^2\theta$ to both sides of our original equation. This gives us $\cot^2\theta + 1 = \csc^2\theta$.

To recap, here are our 3 Pythagorean Identities.

1. $\cos^2\theta + \sin^2\theta = 1$

2. $1 + \tan^2\theta = \sec^2\theta$

3. $\cot^2\theta + 1 = \csc^2\theta$

Now with some more algebraic manipulations we can derive a plethora of trigonometric identities, but this is left to the student as an exercise.

We will return back to some more trigonometric identities later. But for the moment, we will cover two important laws in trigonometry that deal with oblique triangles. They are the Law of Sines and the Law of Cosines. It is important to note that oblique triangles are triangles that do not possess a right angle. These triangles are either acute or obtuse.

LAW OF SINES

The Law of Sines is the first law we will learn with oblique triangles. In order to derive this law, we need to draw an oblique triangle and label some angles and sides. Observe the oblique triangle below.

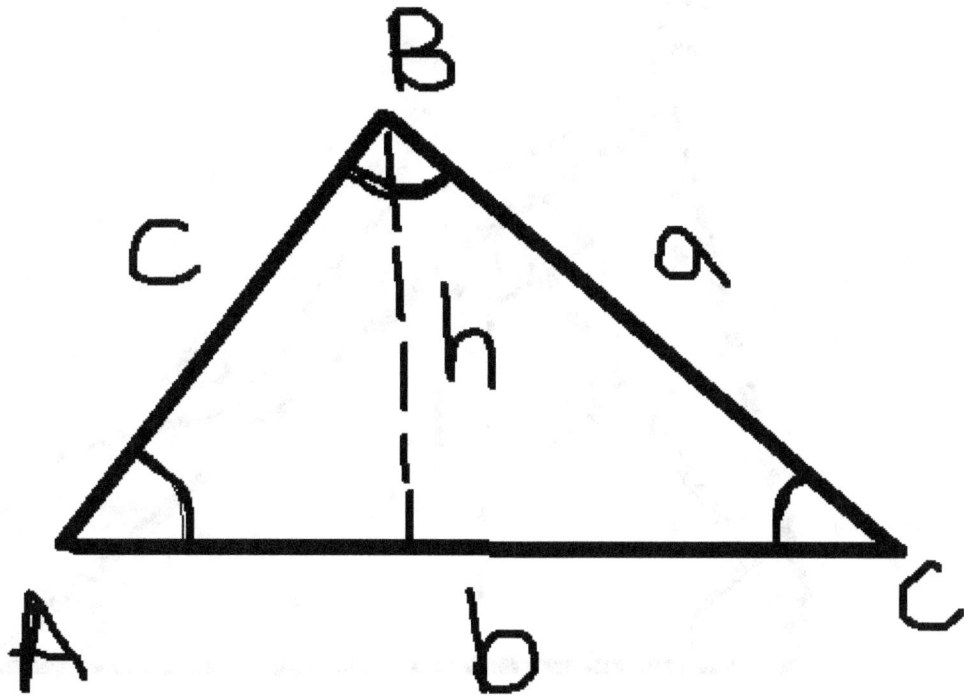

From this drawing we can see that $\sin(A) = h / c$ and $\sin(C) = h / a$. Solving for h in both equations and setting them equal to each other, we see that $c\sin(A) = a\sin(C)$. Now if we divide both sides by $\sin(A)\sin(C)$, we get $a / \sin(A) = c / \sin(C)$.

If we do a further observation of the right triangle and form a right triangle with a leg perpendicular to segment AB and letting the hypotenuse be c, we get another triangle. But notice that this creates two right triangles. See the figure on the next page.

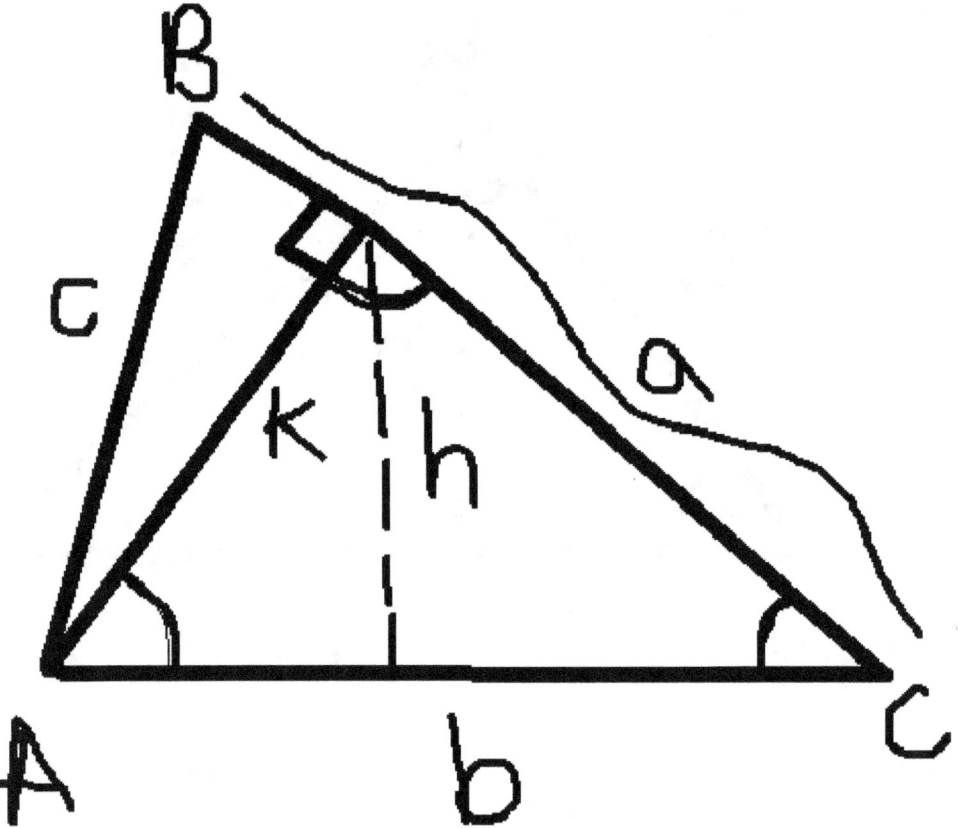

From the new angle B, we have $\sin(B) = k / c$ and $\sin(C) = k / b$. Solving for k in each equation and setting them equal to other we get $b\sin(C) = c\sin(B)$. Then dividing by $\sin(B)\sin(C)$ to both sides of the equation, we have $b / \sin(B) = c / \sin(C)$.

But we know from the first triangle that $a / \sin(A) = c / \sin(C)$, so therefore all three ratios are equal to one another. Here is the Law of Sines.

LAW OF SINES: If triangle ABC is oblique with sides a, b, and c, then

$a / \sin(A) = b / \sin(B) = c / \sin(C)$.

It is important to notice that the Law of Sines can only be used for the following cases that hold.

A. One angle and two sides are known.

B. Two angles and one side are known.

Many mathematicians, engineers, and scientists of the like will agree that the Law of Cosines is an extended version of the Pythagorean Theorem. After deriving the formula, we will see why this is so. We need to observe the triangle on the next page, which is similar to the previous triangle used to derive the Law of Sines.

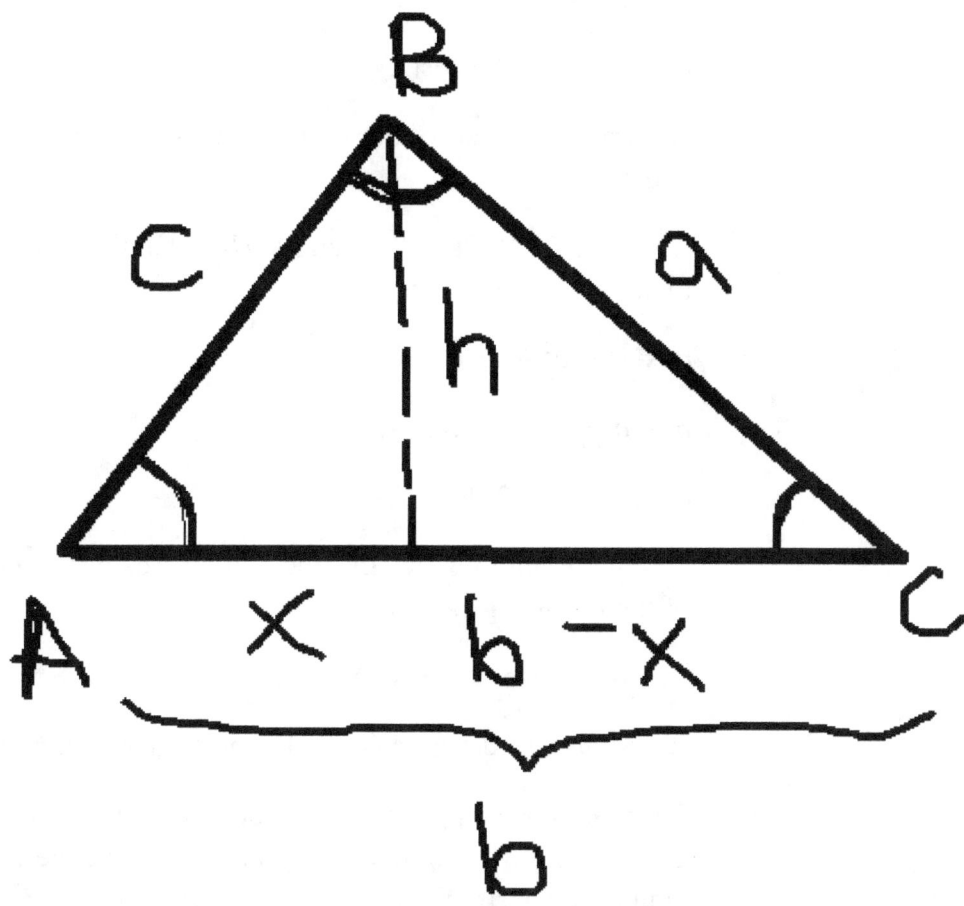

The side b across from angle B is divided into two legs. One leg is length x and the other leg is length $b - x$. If we apply the Pythagorean Theorem to the right triangle on the left, we get $x^2 + h^2 = c^2$. Likewise, if we do this to the right triangle on the right, we get $(b - x)^2 + h^2 = a^2$. Now we expand the left hand side of the second equation.

$(b - x)^2 + h^2 = b^2 - 2bx + x^2 + h^2 = a^2$.

But $x^2 + h^2 = c^2$.

After substitution, we get

$$a^2 = b^2 + c^2 - 2bx.$$

Now we need to express x in terms of cosine. From the right triangle on the left, we see that $\cos(A) = x / c$, which implies that $x = \cos(A)$.

By plugging this value into our equation, we have the first part of the Law of Cosines.

$$a^2 = b^2 + c^2 - 2bc\cos(A)$$

By similar analysis of our oblique triangle, we can find the other two equations.

LAW OF COSINES: If triangle ABC is oblique with sides a, b, and c, then

$$a^2 = b^2 + c^2 - 2bc\cos(A)$$

$$b^2 = a^2 + c^2 - 2ac\cos(B)$$

$$c^2 = a^2 + b^2 - 2ab\cos(C)$$

It is important to notice that the Law of Cosines can only be applied if any of the following cases holds.

A. An angle and two sides are known.

B. All three sides are known, but no angles are known.

There are many applications to both the Law of Sines and the Law of Cosines in physics, engineering, and other sciences. Any situation that requires a construction of an oblique triangle will definitely welcome these two laws to get the job done. We will see how the Law of Cosines can be applied to derive one of the trigonometric identities.

SUM AND DIFFERENCE IDENTITIES

The sum and difference identities can be derived from basic principles of geometry. The first identity we will formulate is the $\cos(x - y)$. But first, we need to take a look at the construction below.

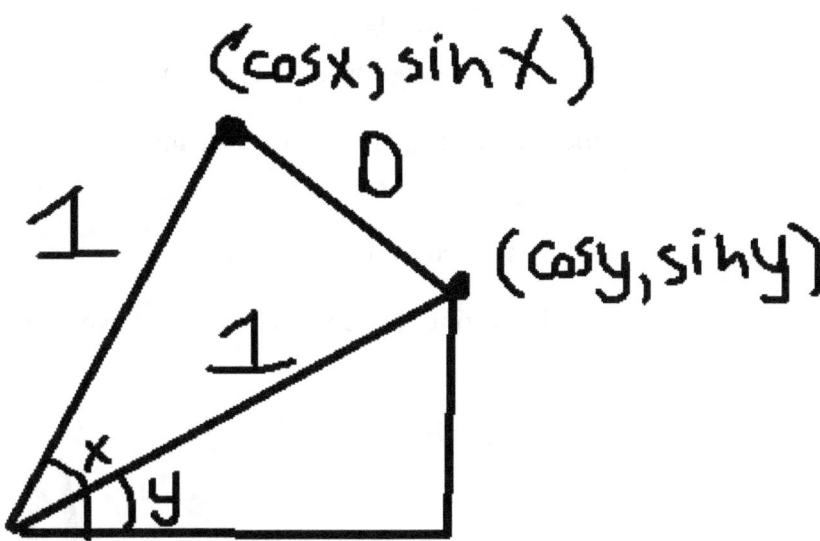

We can apply the distance formula to find the length of side D in terms of sine and cosine. In doing this, we decline to take the square root for reasons that will become quickly apparent.

$$D^2 = (\cos x - \cos y)^2 + (\sin x - \sin y)^2$$

By expanding the binomials on the right side, we have

$$D^2 = \cos^2 x - 2\cos x\cos y + \cos^2 y + \sin^2 x - 2\sin x\sin y + \sin^2 y.$$

But we know from the Pythagorean Identity that $\cos^2 x + \sin^2 x = 1$ and $\cos^2 y + \sin^2 y = 1$.

So, this simplifies to

$$D^2 = 1 - 2\cos x\cos y - 2\sin x\sin y + 1 = 2 - 2\cos x\cos y - 2\sin x\sin y.$$

Obtaining this is half the battle. Now we apply the Law of Cosines to arrive to another equation for D^2.

The angle across from side D is $x - y$. We want to find $\cos(x - y)$.

$$D^2 = 1^2 + 1^2 - 2(1)(1)\cos(x - y) = 2 - 2\cos(x - y).$$

113

Setting both equations equal to each other, we have

$-2\cos(x-y) = 2 - 2\cos x\cos y - 2\sin x\sin y$. Subtracting 2 to both sides and dividing by -2 to both sides we get the identity.

$\cos(x-y) = \cos x\cos y + \sin x\sin y$.

To get the other identity $\cos(x+y)$, we note that $x + y = x - (-y)$. So, we have $\cos(x+y) = \cos(x-(-y)) = \cos x\cos(-y) + \sin x\sin(-y)$.

Since cosine is an even function and sine is an odd function we have the identity.

$\cos(x+y) = \cos x\cos y - \sin x\sin y$

In order to find the identity $\sin(x+y)$, we need to observe the construction below.

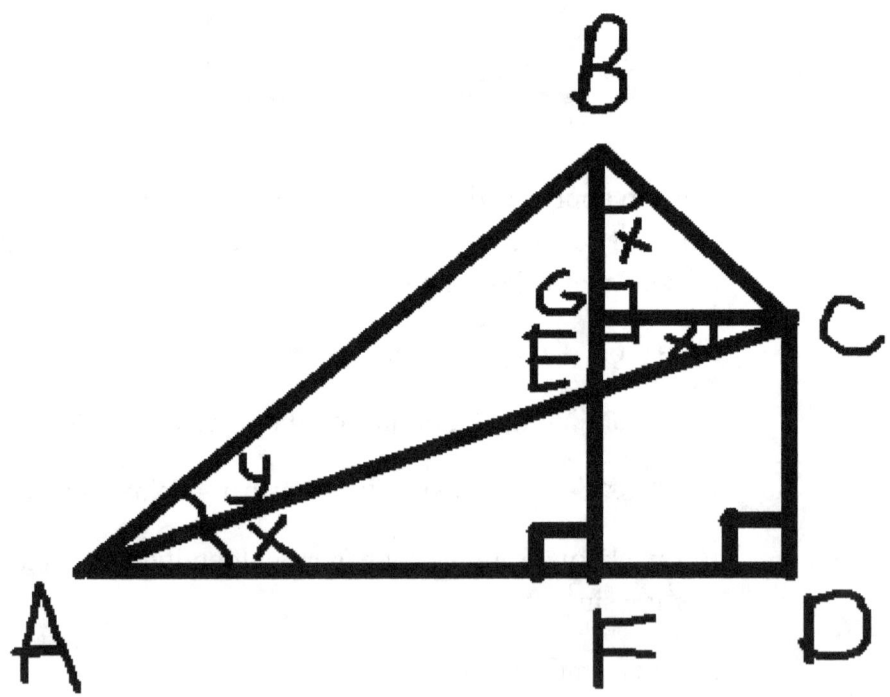

Notice this construction is broken up into a few right triangles. We see that angle $A = x + y$, so our goal is to find $\sin(A)$ or $\sin(x + y)$. Using geometric ratios, we have $\sin(x + y) =$ BF / AB. Now BF is broken into two line segments BG and GF. So, BF = BG + GF. By vertical angles, we see that angles *AEF* and *CEG* are congruent, which are $90 - x$. So, angle A is congruent to angle C. By similar reasoning, we observe that angle B has the same measurement as angles A and C.

Now we have the following.

$\sin(x + y) =$ BF / AB = (BG + GF) / AB = BG / AB + GF / AB.

But GF = CD, so by substitution we have

BG / AB + GF / AB = BG / AB + CD / AB.

Now we apply an algebraic trick to the mix by multiplying top and bottom by BC to the first ratio and multiplying top and bottom by AC to the second ratio to get

BG / AB * BC / BC + CD / AB * AC / AC

Rearranging the terms gives us

BG / BC * BC / AB + CD / AC * AC / AB.

Note that $\cos x =$ BG / BC, $\cos y =$ AC / AB, $\sin x =$ CD / AC, and $\sin y =$ BC / AB.

So, we have $\cos x \sin y + \sin x \cos y$.

By the commutative property, we obtain $\sin(x + y) = \sin x \cos y + \cos x \sin y$.

To get the difference $\sin(x - y)$, we replace y with $-y$ to get $\sin x \cos(-y) + \cos x \sin(-y) = \sin x \cos y - \cos x \sin y$.

The last function we need to obtain for the sum and difference identities is tangent. Proof of these identities will left as an exercise.

$$\tan(x + y) = \frac{\tan x + \tan y}{1 - \tan x \tan y}$$

$$\tan(x - y) = \frac{\tan x - \tan y}{1 + \tan x \tan y}$$

Other identities that are derived from the sum and difference formulas can also be proven.

$$\cos(2x) = \cos^2 x - \sin^2 x = 2\cos^2 x - 1 = 1 - 2\sin^2 x$$

$$\sin(2x) = 2\sin x \cos x$$

PRODUCT TO SUM IDENTITIES

We can apply the sum and difference formulas to obtain the product to sum identities. Let us say we want to add $\cos(x - y) + \cos(x + y)$. This simplifies to

$\cos x \cos y + \sin x \sin y + \cos x \cos y - \sin x \sin y = 2\cos x \cos y$. Then divide by 2 to both sides to get

$\cos x \cos y = \frac{1}{2}(\cos(x - y) + \cos(x + y))$. This shows us that the product of $\cos x$ and $\cos y$ can be written as a sum of the identities we derived earlier.

By a similar method, we can find the other 3 product to sum identities. The proof of these identities will be left as an exercise.

$$\sin x \cos y = \frac{1}{2}(\sin(x + y) + \sin(x - y))$$

$$\cos x \sin y = \frac{1}{2}(\cos(x + y) - \cos(x - y))$$

$$\sin x \sin y = \frac{1}{2}(\cos(x - y) - \cos(x + y))$$

SUM TO PRODUCT IDENTITIES

Again, we can use the sum and difference formulas to find the sum to product identities. Suppose we want to add $\cos x + \cos y$. It seems that we have approached a dead end, but there is an algebraic trick we can employ.

If we let $x = u + v$ and $y = u - v$, then we can use the sum and difference formulas.

Then $\cos x + \cos y =$

$$\cos(u + v) + \cos(u - v) = \cos u \cos v - \sin u \sin v + \cos u \cos v + \sin u \sin v$$

$$= 2\cos u \cos v.$$

Solving for u and v in terms of x and y, we have $u = \frac{1}{2}(x + y)$ and $v = \frac{1}{2}(x - y)$.

Therefore, $\cos x + \cos y = 2\cos\left(\dfrac{x+y}{2}\right)\cos\left(\dfrac{x-y}{2}\right)$.

By a similar approach, we can obtain the other 3 identities. Proofs will be left as an exercise.

$$\cos x - \cos y = -2\sin\left(\dfrac{x+y}{2}\right)\left(\dfrac{x-y}{2}\right)$$

$$\sin x + \sin y = 2\sin\left(\dfrac{x+y}{2}\right)\cos\left(\dfrac{x-y}{2}\right)$$

$$\sin x - \sin y = 2\cos\left(\dfrac{x+y}{2}\right)\sin\left(\dfrac{x-y}{2}\right)$$

This concludes the second part of the trigonometry review.

COMPUTATIONAL EXERCISES #9

1. Solve for triangle ABC if $A = 45°$, $a = 21$, and $b = 16$.

2. Solve the triangle XYZ if $X = 120°$, $y = 5$, and $z = 10$.

3. Two fire stations are 12 miles apart. Fire station A is west of fire station B. A fire located in the mountains is $30°$ east of north from Fire station A and $40°$ west of north from Fire station B. Which fire station is located closest to the mountain fire?

4. A sailor is located 80 miles north of a lighthouse. To avoid some terrible weather conditions, he travels 35 miles at $10°$ west of south. From this point, how far is the sailor from the lighthouse?

5. Solve each of the following trigonometric equations on the interval $[0, 2\pi]$.

    a) $2\sin(x) - 1 = 0$

    b) $\cos^2(x) = 3/4$

c) $(\sin x + \cos x)^2 - 2 = 0$

d) $3\tan^2(x) - 3 = 2\sqrt{3}\tan(x)$

6. Use the identities to evaluate each of the following trigonometric expressions.

a) $\sin(105°)$

b) $\cos(15°)$

c) $\tan(75°)$

d) $\sin(15°) + \sin(75°)$

e) $\cos(2x)\cos(4x)$

f) $\sin(3x) - \sin(x)$ (Write in terms of sin).

g) $\tan(2x)$

CONCEPTUAL EXERCISES #9

1. True or False: $\sin^2 x + \cos^2 x = 1$ for all angles x.

2. True or False: $\sin x \sec x = \cot x$

3. True or False: $\tan^2 x = \sec^2 x - 1$

4. True or False: It is possible to have more than one triangle or no possible triangles when using the Law of Sines.

5. True or False: The Law of Cosines can only be used when only one angle and two sides are known.

6. True or False: $\cos(90° - \theta) = \sin\theta$ for all angles θ.

7. True or False: $\sin(2x) = \sin x \cos x$

8. Prove each of the following trigonometric identities.

a) $(\cot x)(\sin x) = \cos x$

b) $2\sec^2 t - 1 = \tan^2 t - \sec^2 t$

c) $\dfrac{\sin\theta}{1+\cos\theta} = \csc\theta + \cot\theta$

d) $\dfrac{1}{\sec y + 1} = \cot y(\csc y - \cot y)$

9. Show that $\sin^2 x = \dfrac{1-\cos 2x}{2}$ and $\cos^2 x = \dfrac{1+\cos 2x}{2}$. These identities are known as the "half angle formulas."

10. Prove each of the tangent identities.

a) $\tan(x+y) = \dfrac{\tan x + \tan y}{1 - \tan x \tan y}$

b) $\tan(x-y) = \dfrac{\tan x - \tan y}{1 + \tan x \tan y}$

c) Use the identity in a) to show that

$\tan(a + \pi/4) = \sec(2a) + \tan(2a)$.

11. Prove the remaining three product to sum identities.

a) $\sin x \cos y = \tfrac{1}{2}(\sin(x+y) + \sin(x-y))$

b) $\cos x \sin y = \tfrac{1}{2}(\sin(x+y) - \sin(x-y))$

c) $\sin x \sin y = \tfrac{1}{2}(\cos(x-y) - \cos(x+y))$

12. Prove the remaining three sum to product identities.

a) $\cos x - \cos y = -2\sin\left(\dfrac{x+y}{2}\right)\left(\dfrac{x-y}{2}\right)$

b) $\sin x + \sin y = 2\sin\left(\dfrac{x+y}{2}\right)\cos\left(\dfrac{x-y}{2}\right)$

c) $\sin x - \sin y = 2\cos\left(\dfrac{x+y}{2}\right)\sin\left(\dfrac{x-y}{2}\right)$

13. Prove or disprove. $\tan(a+b) - \tan(a-b) = \dfrac{2\tan b}{1 - \tan a \tan b}$

14. Prove or disprove. $\cos 4\theta = 1 - 8(\sin\theta\cos\theta)^2$

15. If $A = \begin{pmatrix} \cos\theta & \sin\theta \\ -\sin\theta & \cos\theta \end{pmatrix}$ and $B = \begin{pmatrix} \cos\phi & -\sin\phi \\ \sin\phi & \cos\phi \end{pmatrix}$, then find each of the following.

a) Show that $AB = BA = \begin{pmatrix} \cos(\theta - \phi) & \sin(\theta - \phi) \\ -\sin(\theta - \phi) & \cos(\theta - \phi) \end{pmatrix}$

b) Prove or disprove. $A^n = \begin{pmatrix} \cos n\theta & \sin n\theta \\ -\sin n\theta & \cos n\theta \end{pmatrix}$

REVIEW #10

TRIGONOMETRY PART 3

The third and final part of the trigonometry review consists of the study of vectors, polar coordinates, and complex numbers in polar form. These topics are essential for more advanced studies in calculus and beyond. While some of these mathematical topics such as vectors overlap in the field of physics, we will do our best to provide both true mathematical and physical insight to help us achieve a clear understanding of the material.

Every student who has taken a course or two in physics knows what a vector is. A vector is both a mathematical and physical quantity that contains both a magnitude and a direction. When we say magnitude, we mean the length or strength of the vector. The direction indicates where the vector is pointing to. Some examples of vectors are velocity, acceleration, and force.

Like matrices, we can perform operations on vectors.

OPERATIONS ON VECTORS

There are a few possible operations we can perform on vectors.

1. Multiply a vector by a scalar. That is if v is a vector and c is any real number, then cv is a vector that is a scalar multiple of v. Note that the word "scalar" is another word for number.

2. Add two vectors v and w together to get $v + w$.

3. Subtract two vectors v and w together to get $v - w$.
   Note that $v - w = v + (-w)$, which is adding the opposite of vector w to vector v.

4. Do a combination of the first three operations listed above. We can add and subtract more than two vectors at a time.

   The question is, how do we perform these operations in a mathematical and physical way? Let us define the mathematical operations first. For simplicity, we will look at vectors in the xy-plane (2-dimensional vectors).

   Let $v = <v_x, v_y>$, $w = <w_x, w_y>$, and c is any real scalar. The v_x, v_y are the x and y components of vector v and w_x, w_y are the x and y components of vector w.

   a) $cv = <cv_x, cv_y>$ We multiply c to both the x and y components of vector v.

   b) $v + w = <v_x + w_x, v_y + w_y>$ We add the corresponding x and y components of vectors v and w together.

   c) $v - w = <v_x - w_x, v_y - w_y>$ We subtract the corresponding x and y components of vectors v and w together.

   There are other properties we can list here such as commutative, associative, and distributive for both addition and multiplication, but we will leave some of these rules for the student to prove as exercises.

   Below are the physical descriptions for a), b), and c) respectively.

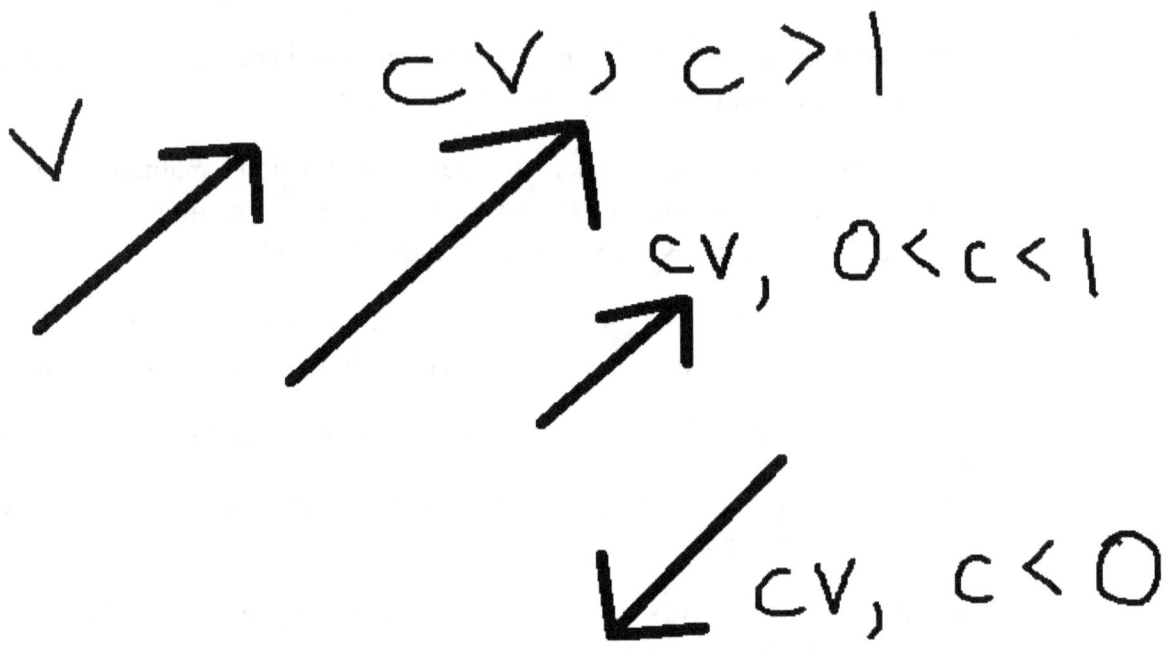

If $c > 1$, then cv grows by a factor of c in the direction of v.

If $0 < c < 1$, then cv shrinks by a factor of c in the direction of v.

If $c < 0$, then cv points in the opposite direction of v and can either grow or shrink. If $|c| > 1$, then the vector grows. If $0 < |c| < 1$, then the vector shrinks.

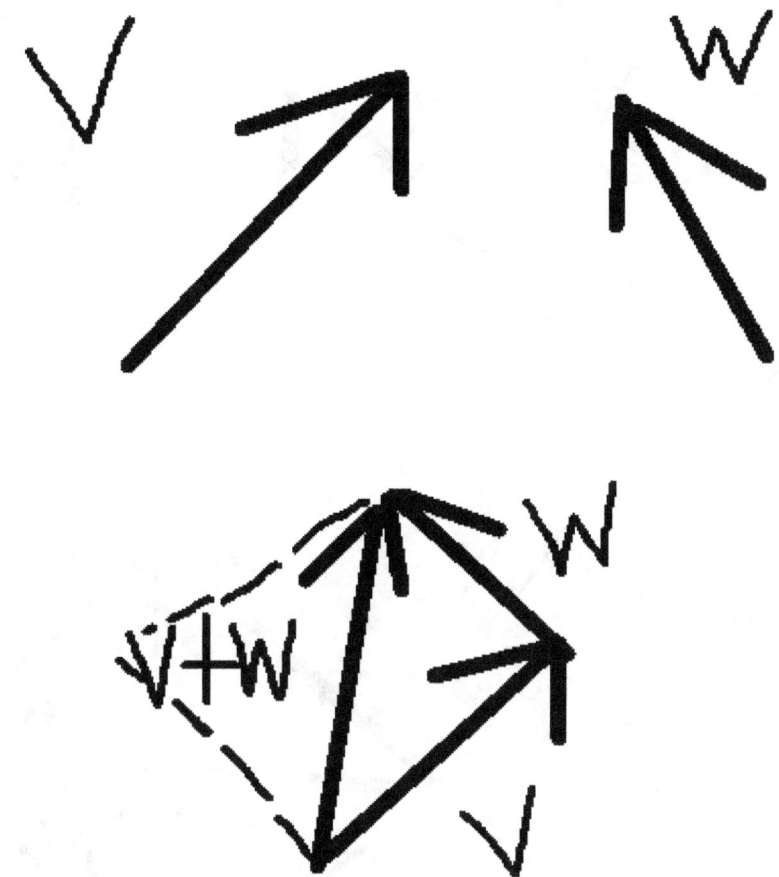

To add vectors *v* and *w* together, we take the head of vector *v* and connect it to the tail of vector *w*. Notice that the head of the resultant vector *v* + *w* is connected to the head of *w*. And the tail of *v* + *w* is connected to the tail of *v*. Also note the horizontal and vertical dashed lines that were drawn to create a parallelogram. The horizontal dashed lines are parallel to *w* with equal length and the vertical dashed lines are parallel to *v* with equal length. We determine the length of the resultant vector by drawing the arrow along the diagonal of the parallelogram. This is known as the "Parallelogram Law" for vectors.

The next description for subtraction is very similar, expect we add the opposite of the second vector to the first vector. So, our resultant vector will look different.

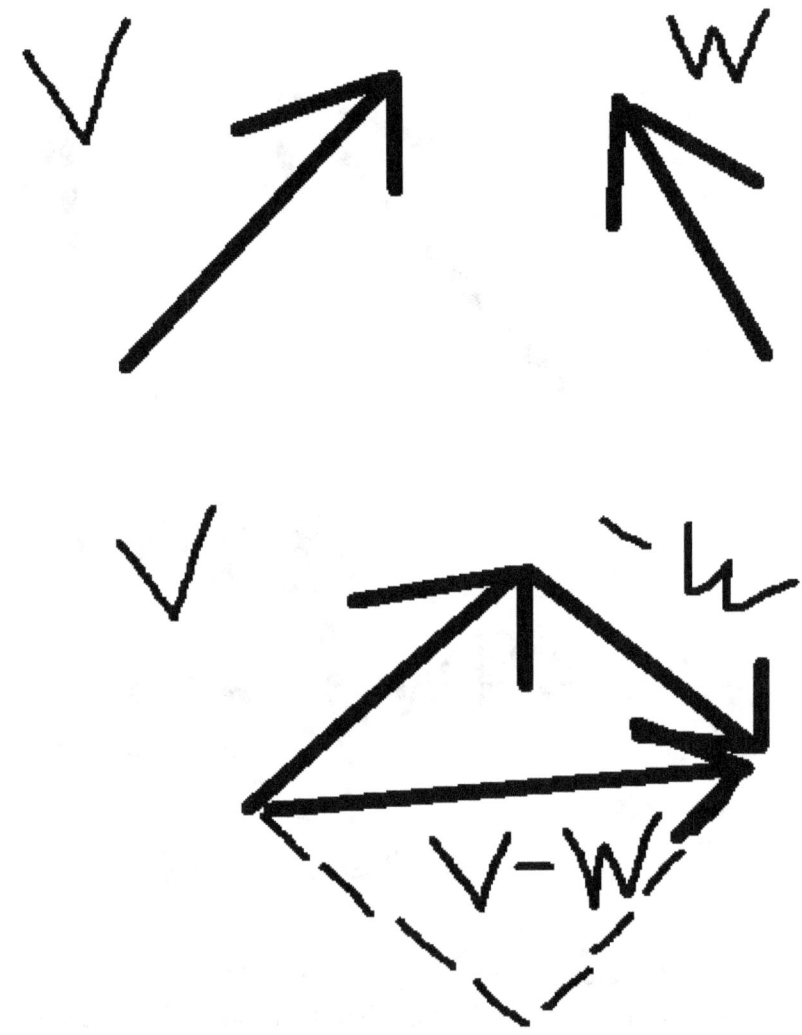

Like in b), we use the same vectors *v* and *w* for subtraction. Notice that the vector −*w* points in the opposite direction of *w*. This is like adding *v* to the opposite of *w*. We apply the Parallelogram Law once more to get the vector *v* − *w*.

We have seen how vectors look in both a mathematical and physical sense. Now it is time to look at some other important properties of vectors that will be very useful.

MAGNITUDE AND DIRECTION OF VECTORS

Recall earlier that the magnitude of a vector is the length or size of the vector. If v is a vector then the magnitude of v is denoted as $\|v\|$. How do we find the magnitude? Here is the definition.

MAGNITUDE OF A VECTOR: If $v = <v_x, v_y>$, then the magnitude of v is defined as $\|v\| = \sqrt{v_x^2 + v_y^2}$.

We see that the magnitude of a vector is an application to the Pythagorean Theorem. We sum up the squares of the x and y components of v and take the square root.

The definition of magnitude can be extended from 3-dimensional vectors up to n-dimensional vectors. All we do is sum up all the squares of the components and take the square root.

How do we find the direction of where the vector is heading? When we talk about direction of a vector, we emphasize where the vector is pointing to and the angle it forms respect to a given axis. Direction is very important when navigators have to read and understand bearings, which are angles formed respect to the north and south axis. We may hear something like the bearing of a ship in the ocean traveling in a direction 15° SE (southeast or east of south).

The next page shows a physical description of the magnitude and direction of a vector.

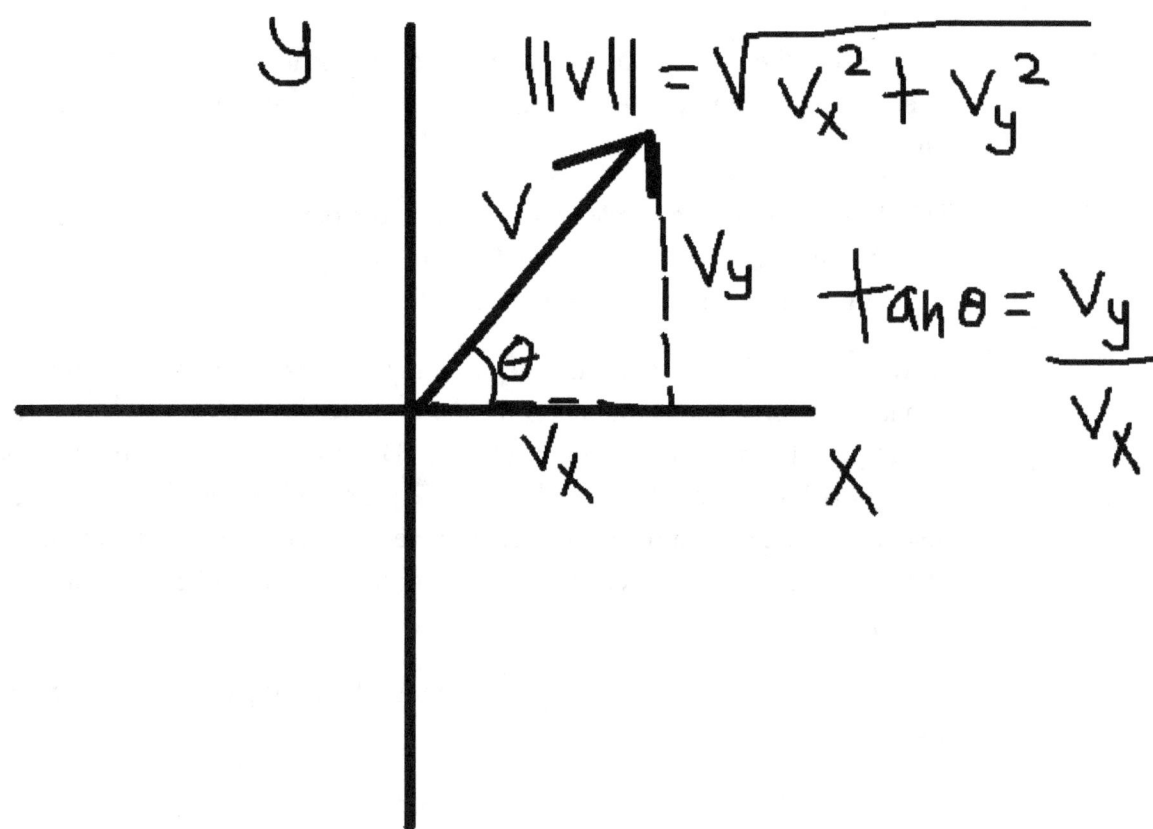

We notice that v is divided up into the x and y components since it is formed at angle θ respect to the x-axis. From trigonometry, we know that $\tan\theta = $ opp / adj.

So, the direction is given as $\tan\theta = v_y / v_x$, which implies that $\theta = \tan^{-1}(v_y / v_x)$.

THE UNIT VECTOR AND NOTATION

Now that we have established a good base on the topic of vectors, we can now discuss the importance of a unit vector and how we use this to express a vector in a different form.

The unit vector can be defined as a vector v divided by its own magnitude or $v / \|v\|$. This vector points in the direction of v and has a magnitude of 1.

Note that the word "unit" in mathematics refers to anything with the number 1.

The student will be asked to prove that the unit vector has a length of 1 unit as an exercise.

In the xy-plane, there are two important unit vectors we will use. They are the i and j vectors. The i unit vector lies on the positive x-axis and the j vector lies on the positive y-axis. Note that i = <1, 0> and j = <0, 1>.

Earlier, we wrote the vector v in terms of components inside the angle brackets. "< >" But we can rewrite v in terms of the unit vectors i and j.

Therefore, $v = <v_x, v_y>$ can be expressed as $v = v_x i + v_y j$.

Now we can even take this expression a step further and define v in terms of its magnitude and angle θ.

From the physical description above, we see that $\cos\theta = v_x / \|v\|$ and $\sin\theta = v_y / \|v\|$. Solving for each of the components we get $v_x = \|v\|\cos\theta$ and $v_y = \|v\|\sin\theta$.

Thus, we can write the vector as $v = \|v\|\cos\theta i + \|v\|\sin\theta j$. If we are given the magnitude of a vector and the angle (direction) of where it points to, then we can acquire each of the individual components.

THE DOT PRODUCT

The dot product is a mathematical operator on two vectors that yields a scalar. Since the result is a scalar, the dot product can be also known as a "scalar product." Below is the definition of dot product.

DEFINITION OF DOT PRODUCT: If $v = <v_1, v_2>$ and $w = <w_1, w_2>$, then we define the dot product of v and w as $v \bullet w = v_1 w_1 + v_2 w_2$.

Note that the left side of the equation can be read as "v dot w." The black dot represents the dot operator performed on v and w.

There are some other important properties that follow from the definition of the dot product. Proofs of these properties can be left as exercises for the student.

PROPERTIES OF DOT PRODUCT

a) $v \bullet v = \| v \|^2$ Magnitude Squared Property

b) $v \bullet w = w \bullet v$ Commutative Property

c) $v \bullet (w + z) = v \bullet w + v \bullet z$ Distributive Property

d) $(cv) \bullet (dw) = (cd)(v \bullet w)$ Scalar Property

e) $v \bullet (rw + z) = r(v \bullet w) + (v \bullet z)$ Bilinear Property

What is the dot product of two non-zero vectors that are perpendicular to one another? Another word for perpendicular is "orthogonal."

DOT PRODUCT OF ORTHOGONAL VECTORS: Two non-zero vectors v and w are orthogonal if and only if $v \bullet w = 0$.

A question of interest is can we express the dot product in terms of the magnitude of vectors v and w and angle θ? The answer is yes! We will derive the geometric interpretation of the dot product.

GEOMETRIC INTERPRETATION OF THE DOT PRODUCT

Consider the diagram with vectors v, w, and $v - w$ with angle θ below.

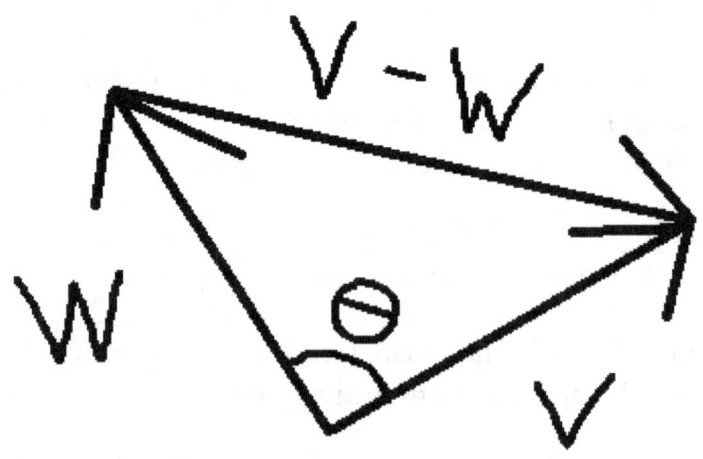

We see the angle θ is formed between vectors v and w, and vector $v - w$ is across from θ. By applying the dot product properties and the Law of Cosines, we can obtain an equation that expresses the dot product in terms of the vectors' magnitudes and the angle between them.

First, we can apply the dot product property a) to vector $v - w$ and then the dot property c).

$$\|v - w\|^2 = (v - w) \bullet (v - w) = (v \bullet v) - (v \bullet w) - (w \bullet v) + (w \bullet w)$$

Rewriting the right hand side of the equation, we get

$$\|v - w\|^2 = \|v\|^2 - 2(v \bullet w) + \|w\|^2.$$

Second, we can apply the Law of Cosines to get

$$\|v - w\|^2 = \|v\|^2 + \|w\|^2 - 2\|v\|\|w\|\cos\theta.$$

If we set the first equation equal to the second equation and do some cancellations, we have the following:

$$v \bullet w = \| v \| \| w \| \cos\theta$$

So, the dot product of vectors v and w is equal to the product of their magnitudes and the cosine of the angle between them.

Some students may be familiar with another vector operator called the "cross product." We will cover this important operation in the Calculus 3 (Vector Calculus) section.

PROJECTION OF A VECTOR ONTO ANOTHER VECTOR

Vector projection is another important operation where we can project one vector onto another vector. That is, if we want to take a vector u and project it onto vector v, the projected vector points in the direction of v. The projected vector is a scalar multiple of v. Observe the diagram below.

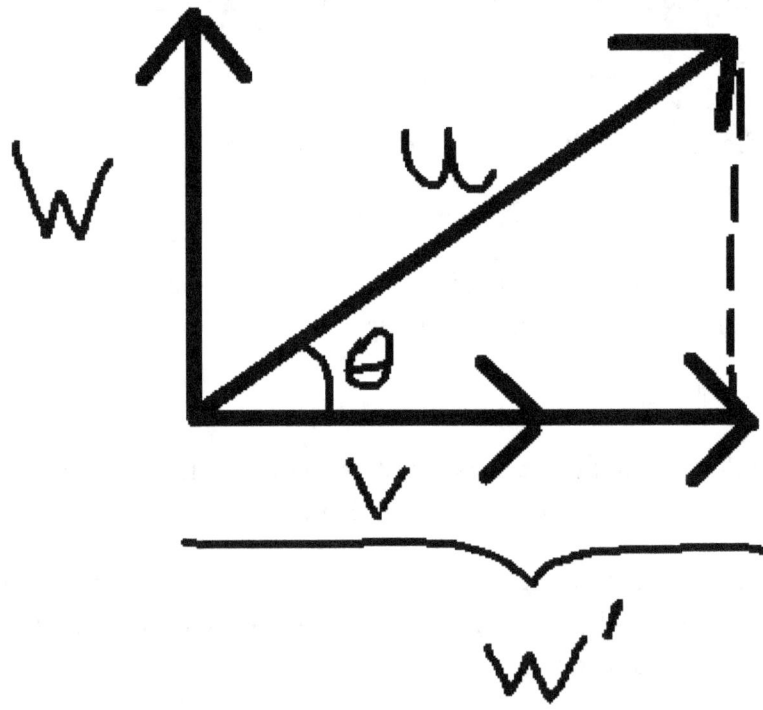

Suppose that w' is the projected vector of u onto v. In shorthand, we can write this as $w' = \text{proj}_v(u)$. We see that $u = w + w'$ and that w' is a scalar multiple of v. So, $w' = cv$ for some positive scalar c. The goal is to find the value of c. If we take the dot product of u and v, we get

$$u \bullet v = (w + w') \bullet v = w \bullet v + w' \bullet v.$$

Since vectors v and w are orthogonal, we have $w \bullet v = 0$. So, this reduces to

$u \bullet v = w' \bullet v$. But $w' = cv$, so $u \bullet v = cv \bullet v$ or $u \bullet v = c \| v \|^2$.

Solving for c, we get $c = \dfrac{u \bullet v}{\| v \|^2}$.

Therefore, we have $\text{proj}_v(u) = \left(\dfrac{u \bullet v}{\|v\|^2} \right) v$.

POLAR COORDINATE SYSTEM

The only system we have been acquainted with for graphing any type of equation is the rectangular system, which is formed by two axes x and y and defines a point (x, y) in the xy-plane. Now we will review another important coordinate system called polar coordinates that defines a point in terms of a radius r and angle θ.

For any point (r, θ) in the polar coordinate system, r represents the radius or distance from a fixed point, and θ represents the angle from a fixed direction, which is usually measured in radians (sometimes degrees). The fixed point is known as a "pole" and the ray formed from the pole to the fixed point is known as the "polar axis." A positive angle rotates counterclockwise from the axis. Observe the diagram below that describes the polar coordinate plane.

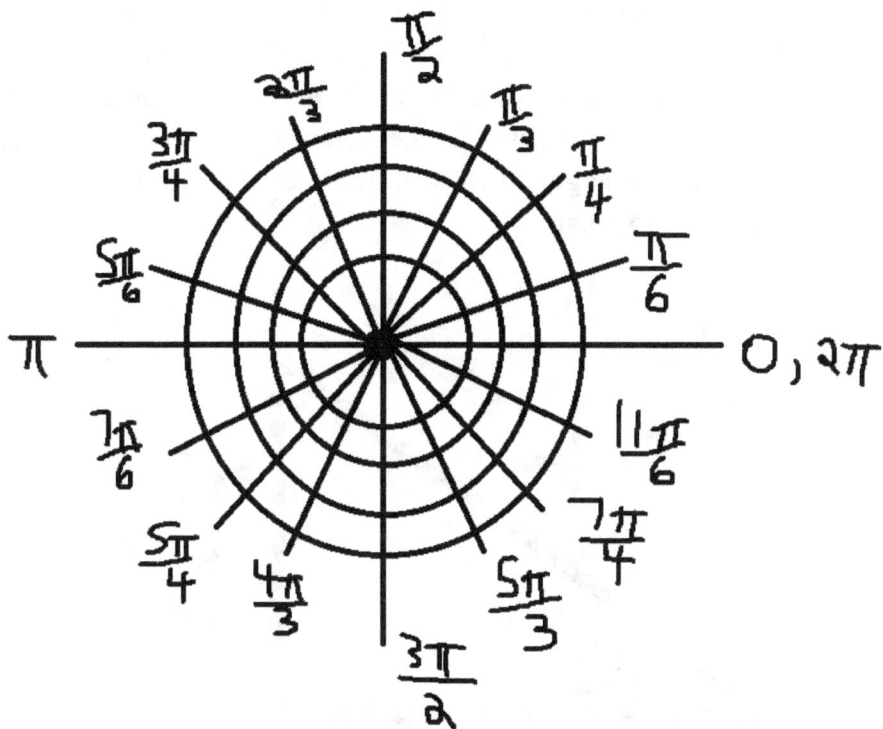

It is important to note that polar coordinates can be expressed an infinite number of ways. That is $(r, \theta) = (r, \theta \pm n \cdot 2\pi) = (-r, \theta \pm (2n+1)\pi)$, where n is any integer.

131

PLOTTING POLAR COORDINATES

The convention of plotting polar coordinates should be fairly simple and straightforward. Let us consider a few cases.

1. If $r > 0$ and $\theta > 0$, then we plot the point r units in the direction of θ from the origin, which is rotated counterclockwise respect to the axis.

2. If $r < 0$ and $\theta > 0$, then we plot the point r units in the opposite direction of θ from the origin, which is rotated counterclockwise respect to the axis.

3. If $r > 0$ and $\theta < 0$, then we plot the point r units in the direction of angle θ, which is rotated clockwise respect to the axis.

4. If $r > 0$, and $\theta < 0$, then we plot the point r units in the opposite direction of angle θ, which is rotated clockwise respect to the axis.

CONVERTING FROM RECTANGULAR TO POLAR

There are times when we want to convert coordinates from one system to another for a number of reasons. Most of these reasons will come in handy when we tackle some calculus problems on the geometric scale. It is important that we master the mechanics first before anything else. When looking at the diagram below, it becomes apparent of how to convert from rectangular to polar.

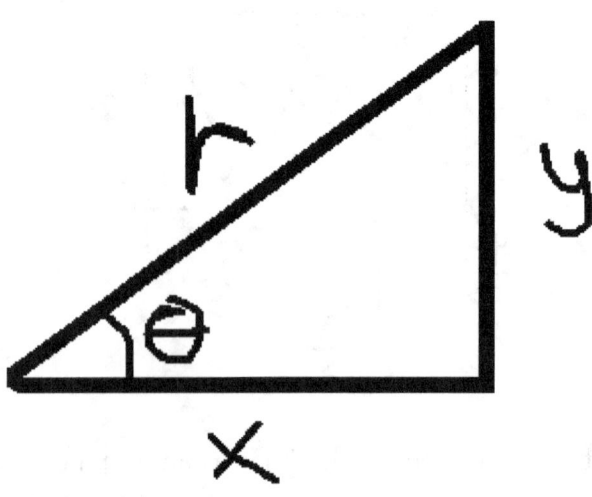

We see the legs of the right triangle are labeled x and y and the hypotenuse is labeled r, which is the radius. The angle θ is formed by the leg x and the hypotenuse r. We can define x and y in terms of both r and θ.

From the analysis of trigonometric functions, we see that

$x = r\cos\theta$, $y = r\sin\theta$, and $\tan\theta = y / x$.

Also note from the Pythagorean Theorem that $x^2 + y^2 = r^2$.

If we have a polar coordinate/equation and want to convert it back to a rectangular coordinate/equation, we can use these identities to perform the conversion.

COMPLEX NUMBERS IN POLAR FORM

Recall in the first review that a complex number is defined by $z = a + bi$, where a is the real part and b is the imaginary part. For conventional purposes, since the real part corresponds to the x-axis and the imaginary part corresponds to the y-axis, we will let $a = x$ and $b = y$. So, the complex number in rectangular form can be written as $z = x + iy$.

If we want to write the complex number in polar form, we use the conversions from the previous section. Since $x = r\cos\theta$ and $y = r\sin\theta$, then we can rewrite the complex number as $z = r\cos\theta + ir\sin\theta$ or $z = r(\cos\theta + i\sin\theta)$.

Now that expression $\cos\theta + i\sin\theta$ is known as Euler's identity. It is written as $e^{i\theta} = \cos\theta + i\sin\theta$. Euler's identity can be proven using calculus and is applied to areas in trigonometry, calculus, and advanced mathematics.

So, another way to write a complex number is $z = re^{i\theta}$.

Another question we want to ask is, can we write the integer power of a complex number using this equation? The answer is yes! It is called de Moivre's Theorem named after the French mathematician Abraham de Moivre.

DE MOIVRE'S THEOREM:

If $z = r(\cos\theta + i\sin\theta)$, then $z^n = r^n(\cos(n\theta) + i\sin(n\theta))$, where n is any integer.

We can also apply De Moivre's Theorem and find the nth roots of any complex number.

NTH ROOTS OF A COMPLEX NUMBER:

If $z = r(\cos\theta + i\sin\theta)$, then $z^{1/n} = r^{1/n}\left(\cos\left(\dfrac{\theta + 2\pi k}{n}\right) + i\sin\left(\dfrac{\theta + 2\pi k}{n}\right)\right)$, where n is a natural number and $k = 0, 1, 2,\ldots n - 1$.

This wraps up the final part of the trigonometry review.

COMPUTATIONAL EXERCISES #10

1. Let $u = \langle 2, -5\rangle$, $v = \langle -6, 10\rangle$, and $w = \langle -3, -8\rangle$. Compute each of the following.

    a) $2u$

    b) $3v - w$

    c) $v - 4u + 5w$

2. Find the magnitude and direction of $v = -6\mathbf{i} + 8\mathbf{j}$. Graph the vector v.

3. Find the angle between the vectors $v = \langle 15, 12\rangle$ and $w = \langle -4, 3\rangle$ in degrees. Round to three decimal places.

4. If $\|u\| = 2$ and $\theta = \pi/6$, write the vector u in $u = x\mathbf{i} + y\mathbf{j}$ form.

5. Determine if $v = \langle 1, -2\rangle$ and $w = \langle -2, -1\rangle$ are orthogonal.

6. Find a unit vector v' that points in the direction of $v = -12\mathbf{i} - 5\mathbf{j}$. Verify that $\|v'\| = 1$.

7. Find the value(s) of k that makes $u = \langle k, -3\rangle$ and $v = \langle -2k, 4k\rangle$ orthogonal.

8. If $u = -\mathbf{i} + \mathbf{j}$ and $v = 3\mathbf{i} - 4\mathbf{j}$, find each of the following.

    a) $\|u\|$

    b) $\|v\|$

    c) The angle between u and v

    d) $\text{proj}_v(u)$ and $\|\text{proj}_v(u)\|$

    e) $\text{proj}_u(v)$ and $\|\text{proj}_u(v)\|$

9. Plot each of the polar coordinates.

   a) $(3, \pi / 4)$

   b) $(-1, 2\pi / 3)$

   c) $(2, -\pi / 6)$

   d) $(-5, -\pi / 2)$

   e) $(4, 7\pi / 4)$

10. Convert each polar coordinate to a rectangular coordinate.

    a) $(10, \pi / 6)$

    b) $(-8, 3\pi / 4)$

    c) $(7, -\pi)$

11. Convert each rectangular coordinate to a polar coordinate.

    a) $(1, -1)$

    b) $(\sqrt{3}, 1)$

    c) $(-8, -6)$

12. Convert each of the rectangular equations to polar equations.

    a) $2x + y = 1$

    b) $y = x^2 - 4$

    c) $y - x = 0$

    d) $x^2 + y^2 = 36$

    e) $y = x / (x + 3)$

13. Write each complex number in polar form.

    a) $z = -1 + i$

    b) $z = \sqrt{3} - i$

    c) $z = 5i$

d) $z = -2$

14. Use Euler's Identity to evaluate each of the following.

a) $e^{i\pi/4}$

b) $e^{i5\pi/6}$

c) $7e^{-i\pi/3}$

d) $|12e^{i11\pi/6}|$

15. Find the nth roots of each complex number.

a) Cube roots of 2.

b) Fourth roots of $3 + 3i$.

c) Fifth roots of $-1 - \sqrt{3}$.

d) Sixth roots of 64.

CONCEPTUAL EXERCISES #10

1. True or False: A vector is a quantity that contains both a magnitude and direction.

2. True or False: The magnitude of a vector can be found by taking the square root of the sum of the squares of the individual components.

3. True or False: The dot product of two vectors results in another vector.

4. True or False: Two vectors that are parallel to one another are orthogonal.

5. True or False: A vector divided by its magnitude is a unit vector. The unit vector points in the direction of the original vector and has a length of 1 unit.

6. True or False: The polar equation $r = 4$ represents a line with a slope of 4.

7. True or False: The rectangular equation $y = \sqrt{3}x$ can be written as $\theta = \pi / 3$ in polar form.

8. True or False: The complex number $z = 1 + i$ can be written as $z = 2e^{i\pi/4}$.

9. True or False: De Moivre's Theorem can be proved by the principle of mathematical induction.

10. True or False: To find the nth roots of a complex number, one must use the equation of $z^{1/n}$ with $k = 0, 1, 2,\ldots n$.

11. Suppose that $u = <u_x, u_y>$, $v = <v_x, v_y>$, and $w = <w_x, w_y>$. Prove each of the following statements.

   a) $u + v = v + u$

   b) $u + (v + w) = (u + v) + w$

   c) $v \bullet w = w \bullet v$

   d) $u \bullet (v + w) = u \bullet v + u \bullet w$

12. If $v = a\mathbf{i} + b\mathbf{j}$ and v' is the unit vector that points in the direction of v, then show that $\|v'\| = 1$.

13. Prove the Cauchy-Schwarz Inequality. $|u \bullet v| \leq \| u \| \| v \|$ Which angle between the two vectors u and v guarantees equality?

14. If $v = <a, -b>$ and $w = <b, a>$, show that

   a) $\|v + w\| = \sqrt{2}\,\|v\| = \sqrt{2}\,\|w\|$

   b) v and w are orthogonal

15. Prove or disprove. If $u \bullet v = u \bullet w$ then $v = w$.

16. Prove de Moivre's Theorem for all natural numbers n.

17. Using Euler's identity, prove each of the following statements.

   a) $e^{-i\theta} = \cos\theta - i\sin\theta$

   b) $e^{i(\theta+\phi)} = \cos(\theta + \phi) + i\sin(\theta + \phi)$

   c) $\cos\theta = \dfrac{e^{i\theta} + e^{-i\theta}}{2}$

d) $\sin\theta = \dfrac{e^{i\theta} - e^{-i\theta}}{2i}$

18. Show that $\ln(-1) = i\pi$.

19. Prove or disprove. $(e^{i\theta})^n = \cos(n\theta) - i\sin(n\theta)$.

20. For any complex number z and any natural number n, find $|z^{1/n}|$.

REVIEW #11

CALCULUS 1 REVIEW PART 1

In the first part of the calculus 1 review we will cover the concepts of limits and continuity. We will review important properties of limits and how to apply these properties to evaluate limits of a function as well as the precise definition of a limit and how to use the definition to prove that the limit exists. We will also review the definition of continuity and go over what makes a function continuous at a given point, and examine some types of discontinuities.

LIMITS

The idea of a limit gives us some valuable information about a given function. When we evaluate a function $f(x)$ at $x = a$, we get the y value $f(a)$. But the limit asks about what function value that $f(x)$ approaches as x approaches a. Notice that approach does not mean equal to. We can look at the limit of a function where x approaches a from the left, from the right, or both. The limit of a function as x approaches a from either the left or right (but not both) is known as a "one-sided limit," which is a subtopic of limits we will review in the future. For now, we will give a basic definition of the limit.

LIMIT DEFINITION: Let $f(x)$ be a real function and let L be some real number.

$$\lim_{x \to a} f(x) = L$$

This is read as "The limit of $f(x)$ as x approaches a equals L." When we mean x approaches a, we mean x is getting close to values slightly less than a and values slightly greater than a, but not equal to a.

If these values that x approaches from the left and right side both yield the same function value of L, then we say that the limit exists. Otherwise, the limit does not exist and may denote as D.N.E. = DOES NOT EXIST. Look at the graph below.

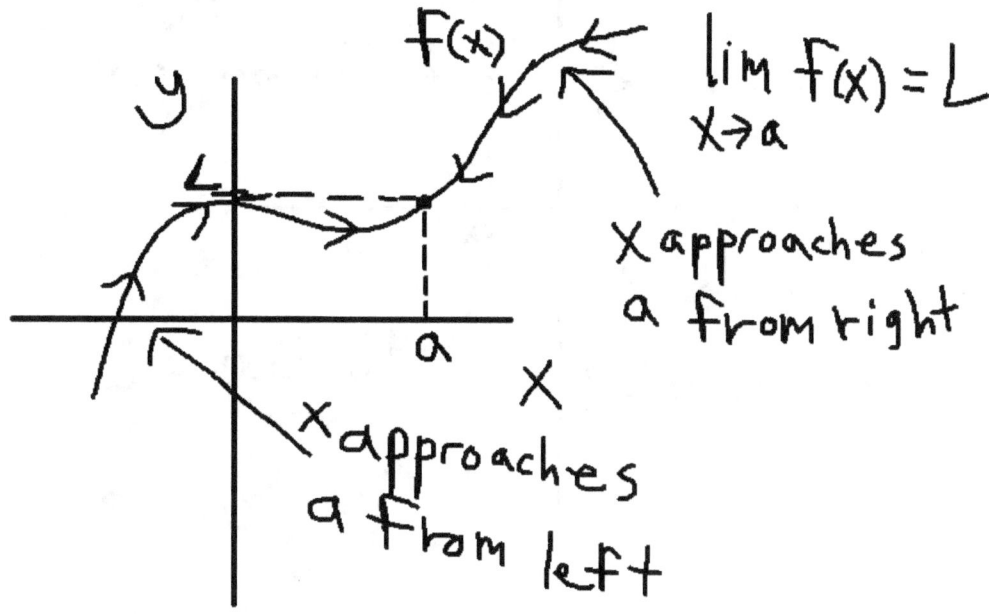

ONE-SIDED LIMITS

When we discuss one-sided limits, there are two different types:

1. Left-Handed Limits and 2. Right-Handed Limits

Left-Handed Limits of a function are where x approaches a from the left, and Right-Handed Limits of a function are where x approaches a from the right. Below are the one-sided limit notations.

LEFT-HANDED LIMIT: $\displaystyle\lim_{x \to a^-} f(x) = L$, where L is a real number. The superscripted hyphen next to the a indicates a left-handed limit.

RIGHT-HANDED LIMIT: $\displaystyle\lim_{x \to a^+} f(x) = R$, where R is a real number. The superscripted plus sign next to the a indicates a right-handed limit.

If $L = R$, then $\displaystyle\lim_{x \to a} f(x)$ exists. Otherwise the limit does not exist. See the graph of $f(x)$ below that demonstrates one-sided limits.

$$\lim_{x \to a^-} f(x) = L$$

$$\lim_{x \to a^+} f(x) = R$$

$$\lim_{x \to a} f(x) \quad \underline{D.N.E.}$$

In this graph, the one-sided limits are unequal, so the limit of $f(x)$ where x approaches a in both directions does not exist. But in the previous graph, both the one-sided limits are equal. So, the limit of $f(x)$ where x approaches a in both directions does exist, and is equal to the left/right-handed limits.

METHODS OF COMPUTING LIMITS

There are a few methods that we can employ to compute limits.

1. NUMERICAL METHOD

2. GRAPHICAL METHOD

3. ANALYTICAL METHOD

The numerical method is more of a mechanical approach, which requires us to plug in values of x that are slightly less than a or slightly greater than a. For example, if we wanted to know the limit of $f(x)$ as $x \to 1$, then we can plug in values slightly less than 1 like 0.9, 0.99, 0.999, etc, and we can plug in values slightly greater than 1 like 1.01, 1.001, 1.0001, etc. If the function value approaches the same number from either side, then we know the limit of that function exists. The numerical method is a good approach for functions that are difficult to graph or simplify analytically in the absence of a calculator or computer.

The graphical method is a visual approach that helps us to see how the function is behaving around the values of a. Looking at the graph on the previous page, we can instantly apply the graphical method and determine one-sided and general limits with minimal effort.

The analytical method can be used when functions can be simplified by applying the rules of algebra such as factoring, cancellations, rewriting, other properties, etc. The analytical method should be the first choice when evaluating limits of elementary functions (algebraic, rational, periodic, and transcendental).

LIMIT PROPERTIES

Below are some important limit properties that every student should know when computing limits.

$$\lim_{x \to a}(cf(x)) = c\lim_{x \to a} f(x) \text{ for any real number } c.$$

$$\lim_{x \to a}(f(x) \pm g(x)) = \lim_{x \to a} f(x) \pm \lim_{x \to a} g(x)$$

$$\lim_{x \to a}[f(x) \cdot g(x)] = \lim_{x \to a} f(x) \cdot \lim_{x \to a} g(x)$$

$$\lim_{x \to a}\left(\frac{f(x)}{g(x)}\right) = \frac{\lim_{x \to a} f(x)}{\lim_{x \to a} g(x)}, \text{ provided that } g(x) \neq 0.$$

$$\lim_{x \to a} c = c$$

$$\lim_{x \to a}[f(x)]^n = \left[\lim_{x \to a} f(x)\right]^n \text{ for any real number } n.$$

Note that the last property also applies to rational exponents.

Another important fact we should know about limits deals with inequalities. Suppose that we have well behaved, real functions $f(x)$ and $g(x)$ so that $f(x) \leq g(x)$ for all x in the closed interval $[a, b]$ and $a \leq c \leq b$. The inequality is not necessarily true at $x = c$. Then we have $\lim\limits_{x \to c} f(x) \leq \lim\limits_{x \to c} g(x)$.

This fact can be taken a step further when we introduce a third function into the mix, which leads us to an important theorem of limits. The theorem is known as "squeeze theorem" or sometimes called the "sandwich theorem."

SQUEEZE/SANDWICH THEOREM

For any x (except maybe $x = c$) in $[a, b]$, we have $f(x) \leq h(x) \leq g(x)$.

Assume that $\lim\limits_{x \to c} f(x) = L$ and $\lim\limits_{x \to c} g(x) = L$.

Then $\lim\limits_{x \to c} h(x) = L$.

In order to apply the squeeze theorem, the inequality of the functions f, g, and h must hold true and the limits of both functions f and g must be equal. This theorem tells us that the function h is "squeezed" or "sandwiched" in between the functions f and g and shows us how all three functions behave around $x = c$, but not right at $x = c$. See the graph below that illustrates the squeeze/sandwich theorem.

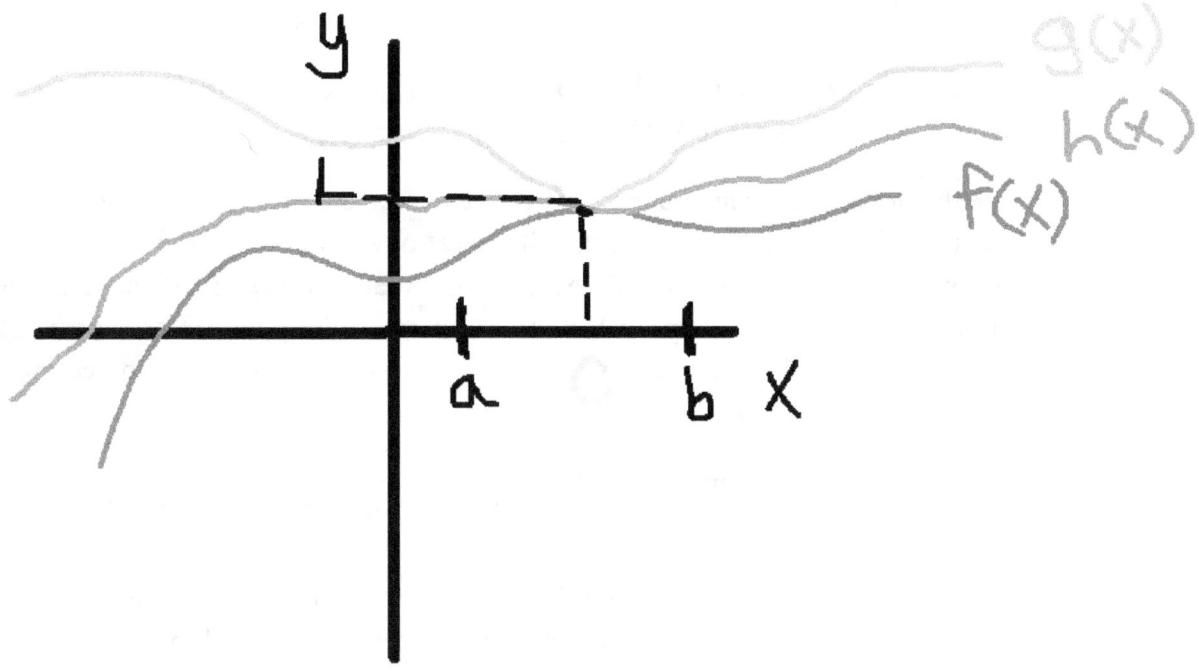

INFINITE LIMITS

Sometimes we want to find the limit of a function as *x* approaches a certain number and we notice that the function value is extremely large or small at that point. This may be due to the function increasing or decreasing without bound, which possesses vertical asymptotes at that point. Here is an important fact involving infinite limits.

If $\lim_{x \to a} f(x) = \pm\infty$, $\lim_{x \to a^-} f(x) = \pm\infty$, and $\lim_{x \to a} f(x) = \pm\infty$, then f(x) has a vertical asymptote at $x = a$.

This fact says that if *f*(*x*) approaches a really large or really small value near $x = a$, then we are guaranteed a vertical asymptote at $x = a$. This kind of behavior is prevalent in rational functions where we have at least one vertical asymptote.

Now if we have two functions $f(x)$ and $g(x)$ such that $\lim_{x \to a} g(x) = \infty$, then

$\lim_{x \to a} \dfrac{f(x)}{g(x)} = 0$. If we have the limit for $f(x)$ go to infinity, then we have ∞ / ∞, which is an indeterminate form. Other indeterminate forms that we cannot evaluate for limits are $0/0$, 0^{∞}, $\infty \pm \infty$, etc.

On the other hand, what happens when we evaluate a limit of a function as x approaches infinity or negative infinity? Rational functions also have horizontal asymptotes at certain values when x becomes really large or really small. The limit of a function as x approaches plus or minus infinity may either approach 0 or plus or minus infinity. These are cases we have to consider when dealing with infinite limits.

If $\lim_{x \to \infty} f(x) = L$ or $\lim_{x \to -\infty} f(x) = L$, then $f(x)$ has a horizontal asymptote at $y = L$.

In Review #2, we discuss the possible cases for horizontal asymptotes of rational numbers. Let us review these again in the context of infinite limits.

If the degree of $f(x)$ and degree of $g(x)$ are equal and the leading coefficient of $f(x)$ is a, and the leading coefficient of $g(x)$ is b, then

$\lim_{x \to \infty} \dfrac{f(x)}{g(x)} = \dfrac{a}{b}$ and $\lim_{x \to -\infty} \dfrac{f(x)}{g(x)} = \dfrac{a}{b}$. Thus, we have a horizontal asymptote at

$y = a / b.$

If the degree of $f(x)$ is one less than the degree of $g(x)$, then the limit goes to zero. Thus, we have a horizontal asymptote at $y = 0$.

If the degree of $f(x)$ is one more than the degree of $g(x)$, then the limit goes to plus or minus infinity and has a slant asymptote in sloe intercept form $y = mx + b.$

If c is a non-zero real number and $\lim_{x \to \infty} f(x) = \pm\infty$ or $\lim_{x \to -\infty} f(x) = \pm\infty$, then

$\lim_{x \to \infty} \dfrac{c}{f(x)} = 0$ and $\lim_{x \to -\infty} \dfrac{c}{f(x)} = 0$.

Now that we have established the basic rules for limits, we need to define some rules for continuity of a function and see how this concept is directly related to limits.

IMPORTANT LIMITS

Here are three important limits that can be very useful in the study of calculus.

1. $\lim_{x \to 0} \dfrac{\sin x}{x} = 1$

2. $\lim_{x \to 0} \dfrac{\cos x - 1}{x} = 0$

3. $\lim_{x \to 0} (1 + x)^{1/x} = e$

CONTINUITY

Continuity is a property of functions that tells us how the function is behaving. We want to determine if a function behaves like a smooth curve or if it has any holes, jumps, asymptotes or other unusual behaviors. A simple, elementary definition of a continuous function is that we are able to trace through the function on paper without picking up our pencil. We will see many functions that are continuous and many functions that are not continuous. But first, let us define the continuity of a function at a certain point.

Let $f(x)$ be a function. There are three conditions that must be satisfied in order for $f(x)$ to be continuous at $x = a$.

1. $f(a)$ is defined.

2. $\lim_{x \to a} f(x)$ exists.

3. $\lim_{x \to a} f(x) = f(a)$

In other words, if any of these three conditions does not hold then the function is not continuous at $x = a$.

For $f(x)$ to be continuous on $[a, b]$, all three conditions of continuity must be satisfied for all x values in that interval.

There are three types of discontinuities that we need to know when a function is not continuous. They are jump, point, and asymptotic discontinuities.

JUMP DISCONTINUITY: A jump discontinuity occurs when one part of the function is not connected to another part of the function. If we were to trace the function on a piece of paper, we would have to "jump" from one piece of the function to the other piece. Jump discontinuities are very common in piecewise functions. See the graph below.

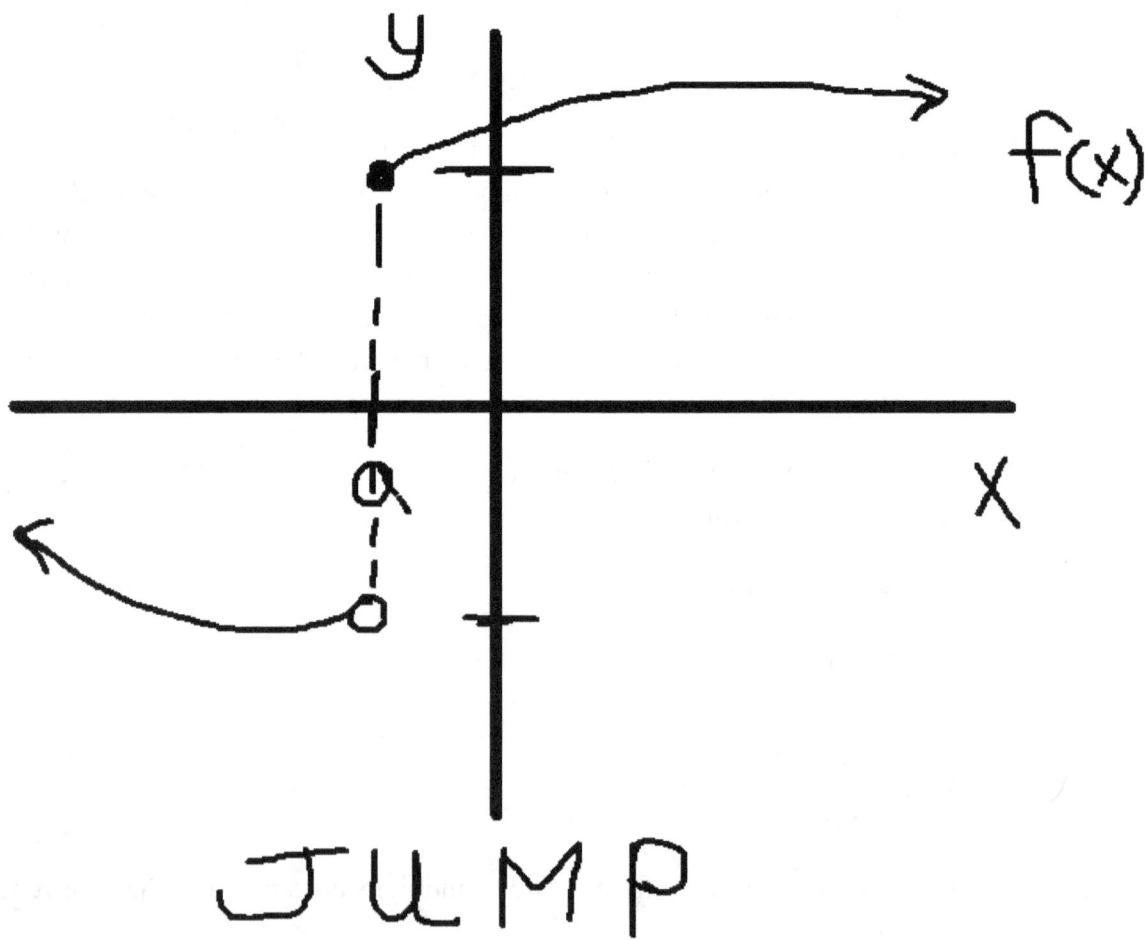

POINT DISCONTINUITY: A point discontinuity is when we have a function (mostly piecewise) that can be defined for more than one value at a certain point. Since the function has more than one value at that point, then we know the function is not continuous. If we can redefine the function, so we can "fill in" the hole or gap, then we have a "removable discontinuity." As a result, the function is now continuous at that point. See the graph below.

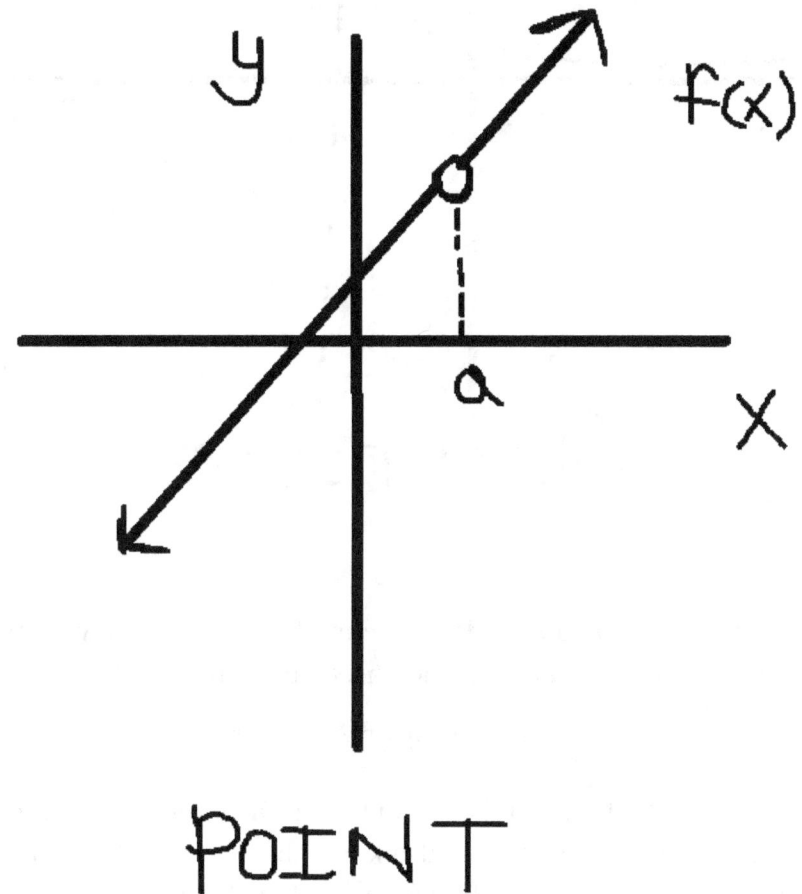

ASYMPTOTIC DISCONTINUITY: Asymptotic discontinuity occurs for functions that have vertical asymptotes. Rational functions and some trigonometric functions possess this type of discontinuity. We can see that the function is not fully connected since the dotted line of the vertical asymptote divides the function into two parts. See the graph below.

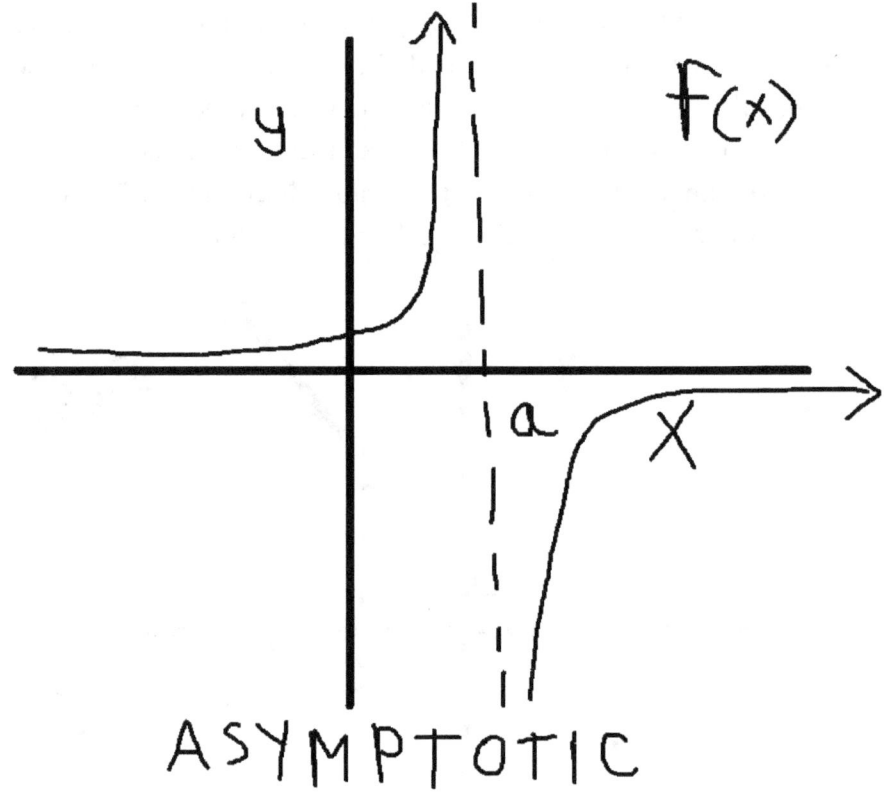

y

f(x)

a

X

ASYMPTOTIC

Now that we have a good understanding of continuity, we will move on to an important theorem in calculus known as the Intermediate Value Theorem.

INTERMEDIATE VALUE THEOREM

We learned that a function $f(x)$ is continuous at a given point $x = a$ provided that the limit exists and is $f(a)$. We also know that $f(x)$ is continuous on $[a, b]$ if it continuous at all values of x in that interval. From this, we can establish the Intermediate Value Theorem.

INTERMEDIATE VALUE THEOREM: Let $f(x)$ be a real, continuous function on $[a, b]$ and let R be any number between $f(a)$ and $f(b)$. Then there exists a number c such $a < c < b$ and $f(c) = R$.

This means that we can pick any value R between the values $f(a)$ and $f(b)$ and find a number c between a and b so that the function value evaluated at $x = c$ is R.

Consequently, the Intermediate Value Theorem can help us resolve the case when $R = 0$, i.e. $f(c) = 0$. This tells us that there exists a root at $x = c$ in between

148

a and *b* since *f(a)* and *f(b)* have opposite signs. The Intermediate Value Theorem does not tell us exactly what that root is, but it guarantees the existence of a root in that interval. See the graphs below.

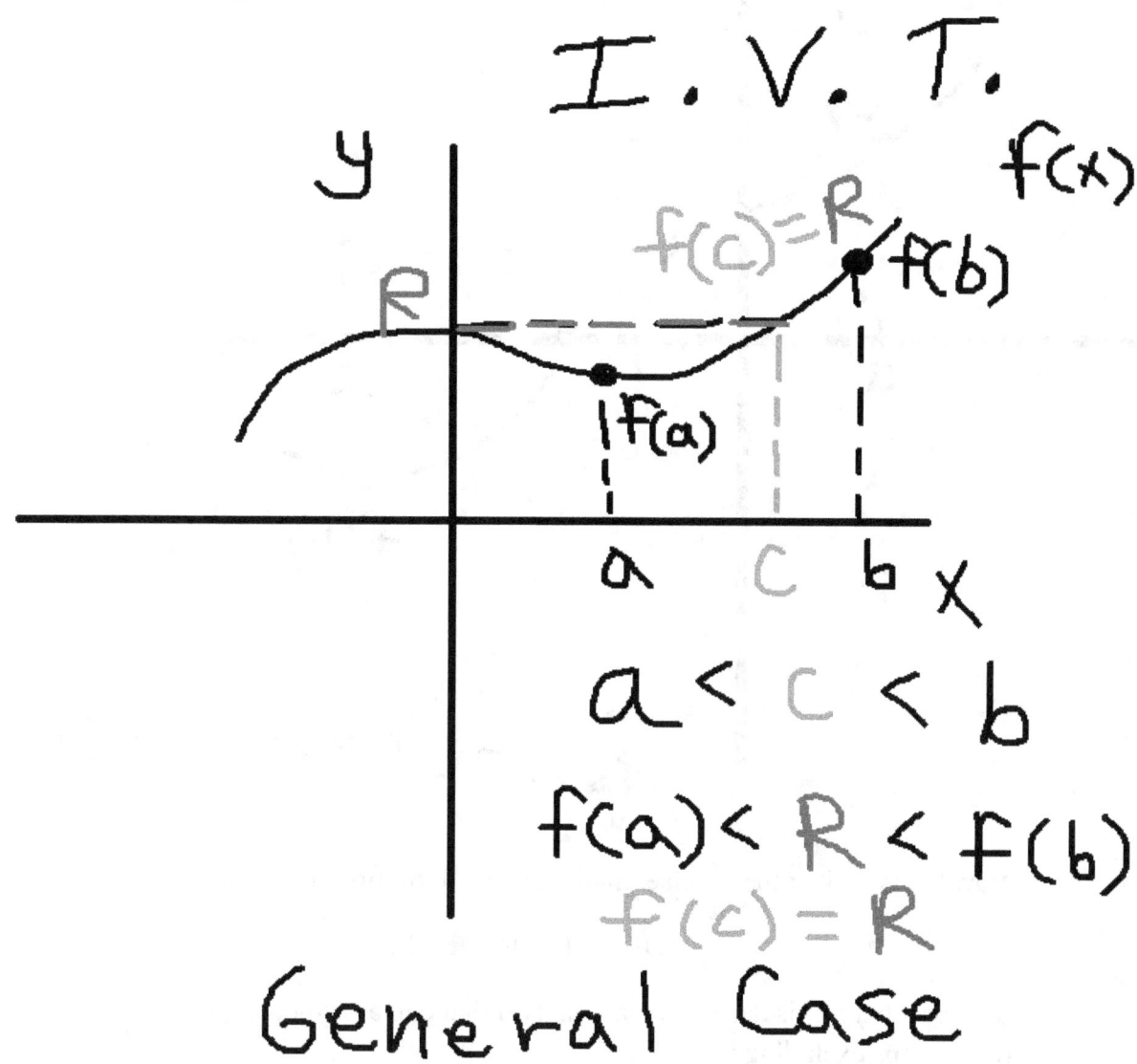

I. V. T.

$$a < c < b$$

$$f(a) < R < f(b)$$

$$f(c) = R$$

General Case

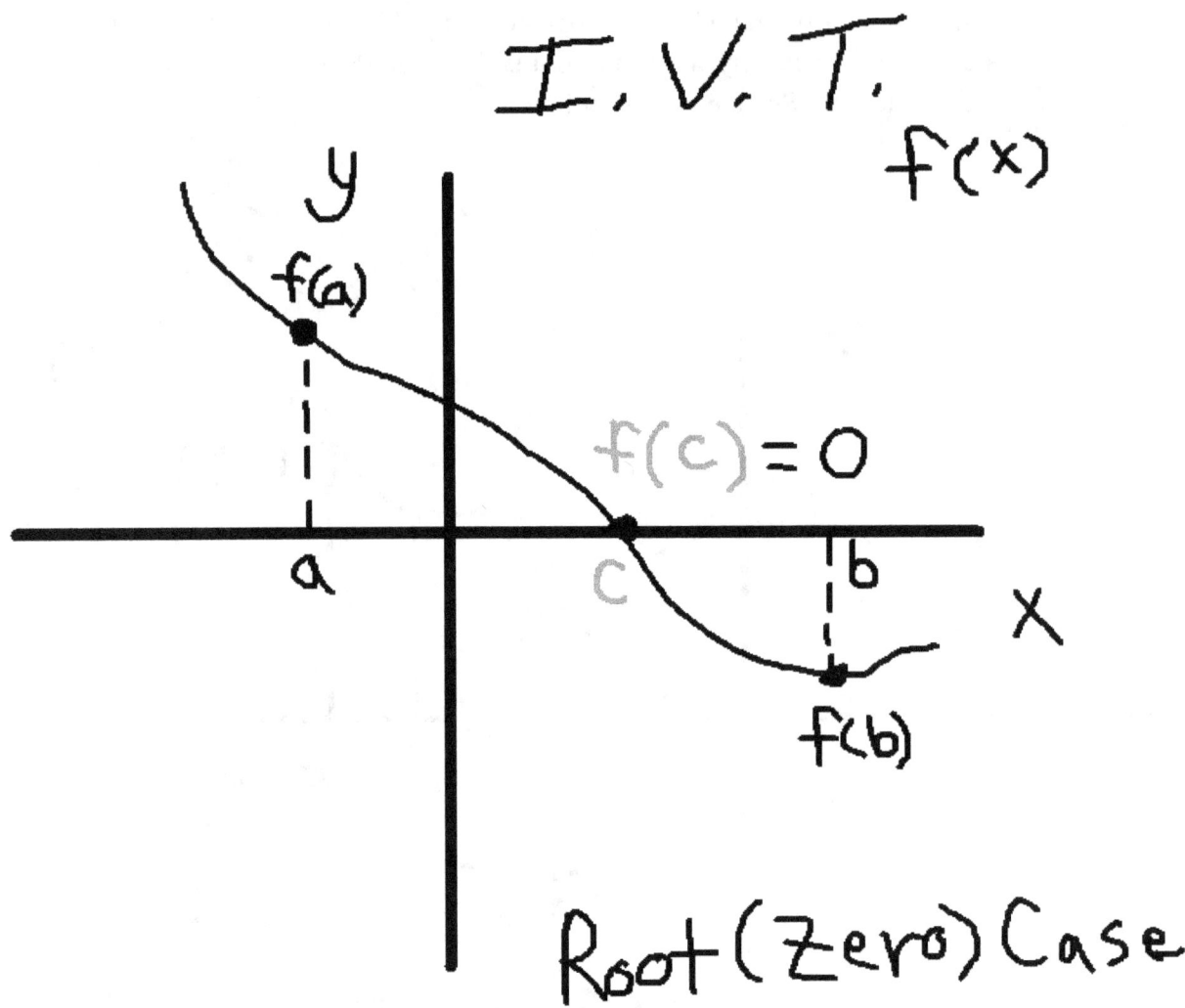

I. V. T.

f(x)

f(c) = 0

Root (Zero) Case

Now let us look at the precise, mathematical definition of a limit.

DEFINITION OF LIMIT

Suppose that $f(x)$ is defined on an interval that contains the value $x = a$, (or perhaps excluding the value $x = a$).

Then $\lim_{x \to a} f(x) = L$ if for every $\varepsilon > 0$, there exists a $\delta > 0$ such that

$0 < | x - a | < \delta$ implies that $| f(x) - L | < \varepsilon$.

This definition seems rather complicated and can be difficult to understand by the notation alone. All this is telling us is that given any positive value of epsilon we can choose a delta small enough so that the distance between the function $f(x)$

and the limit L is less than epsilon. We can view this in a graphical sense by considering the interval $a - \delta < x < a + \delta$ on the x-axis and the interval $L + \varepsilon < L < L - \varepsilon$ on the y-axis. If we draw red, vertical dotted lines for the endpoints of the first interval and purple, horizontal dotted lines for the endpoints of the second interval, we can see an overlapping region between the two intervals, which is highlighted in green. If we are to choose any x in $(a - \delta, a + \delta)$, then we can see that the distance between x and a is less than δ. Likewise, we can see that the distance between $f(x)$ and the limit L is less than ε. See the graph below that illustrates the definition of the limit.

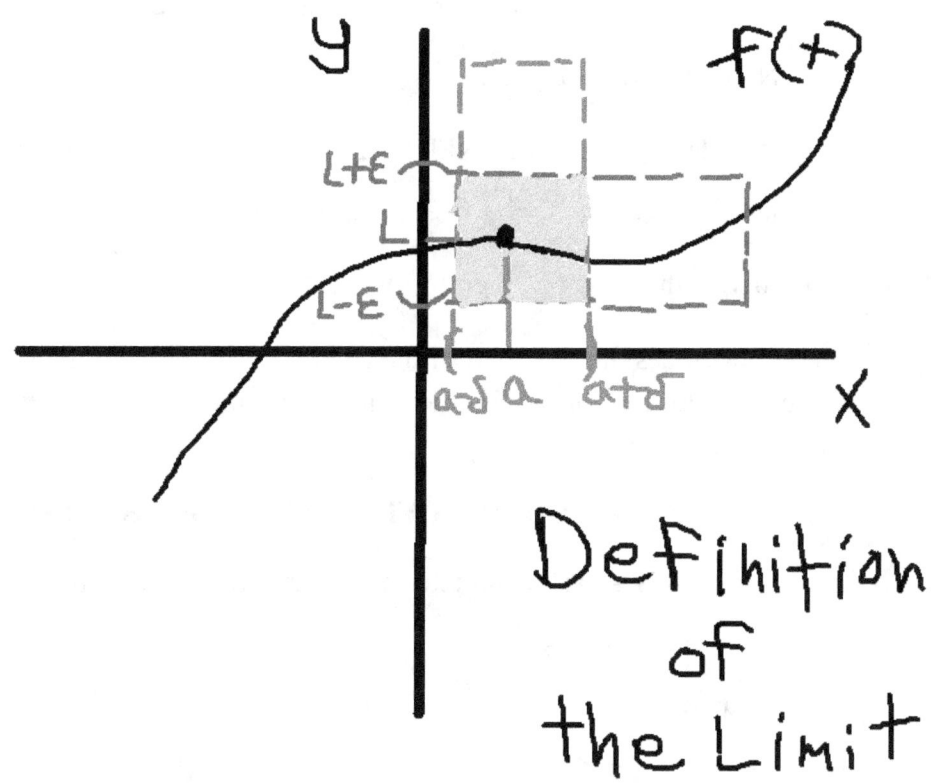

DEFINITIONS OF INFINITE LIMITS AND LIMITS TO INFINITY

If we want to prove that the limit of function $f(x)$ as x approaches a is ∞ or $-\infty$, then we need the following definitions of infinite limits. We will omit illustrations for these definitions.

POSITIVE INFINITY: Let $f(x)$ be defined on an interval that contains $x = a$, (possibly excluding $x = a$). Then $\lim_{x \to a} f(x) = \infty$ if for any real number $N > 0$, there exists a $\delta > 0$ such that if $0 < |x - a| < \delta$ then $f(x) > N$.

NEGATIVE INFINITY: Let $f(x)$ be defined on an interval that contains $x = a$, (possibly excluding $x = a$). Then $\lim_{x \to a} f(x) = -\infty$ if for any real number $N < 0$, there exists a $\delta > 0$ such that if $0 < |x - a| < \delta$ then $f(x) < N$.

LIMITS TO INFINITY: Let $f(x)$ be defined on $x > M$ for some real number M. Then $\lim_{x \to \infty} f(x) = L$ if for any real number $N > 0$, there exists a $\delta > 0$ such that if $x > N$, then $|f(x) - L| < \varepsilon$.

LIMITS TO NEGATIVE INFINITY: Let $f(x)$ be defined on $x < M$ for some real number M. Then $\lim_{x \to -\infty} f(x) = L$ if for any real number $N < 0$, there exists a $\delta > 0$ such that if $x < N$, then $|f(x) - L| < \varepsilon$.

This concludes our first part of the calculus 1 review. In the next review, we will apply the concept of the limit to help us find the slope of the tangent line, which is known as the derivative.

COMPUTATIONAL EXERCISES #11

For exercises 1-10, evaluate each of the following limits.

1. $\lim_{x \to 2} \left(x^2 - 4x + 3 \right)$

2. $\lim_{x \to -1} \dfrac{-6x + 6}{x^2 - 1}$

3. $\lim_{x \to \infty} e^{-x}$

4. $\lim_{x \to -4} \dfrac{3x}{\sqrt{x + 4} - x}$ Hint: Rationalize the denominator.

5. $\lim\limits_{x \to -\infty} \dfrac{2x^3 + 1}{-5x^2 + 10x - 7}$

6. $\lim\limits_{x \to 0} \dfrac{\sin 4x}{x}$

7. $\lim\limits_{x \to \infty} \tan^{-1} x$

8. $\lim\limits_{x \to 2} \dfrac{1}{x - 2}$

9. $\lim\limits_{x \to -\infty} \dfrac{4x^2 - x + 5}{9 - x^2}$

10. $\lim\limits_{x \to 1} \ln(x - 1)$

11. Let $f(x) = [x + 5, \text{ if } x \geq 3$

   $[-2x, \text{ if } x < 3$

   Evaluate each of the following limits.

   a) $\lim\limits_{x \to 3^+} f(x)$

   b) $\lim\limits_{x \to 3^-} f(x)$

   c) $\lim\limits_{x \to 3} f(x)$

12. Find the value of k so that $f(x) = [\dfrac{x^3 + 8}{x + 2}, x \neq 2$ is continuous.

$$[kx^2 - 4, x \geq 2$$

13. Look at the following graph of $f(x)$ below.

   a) Where is $f(x)$ undefined?

   b) For what intervals is $f(x)$ continuous?

   c) For what values of x is $f(x)$ discontinuous?

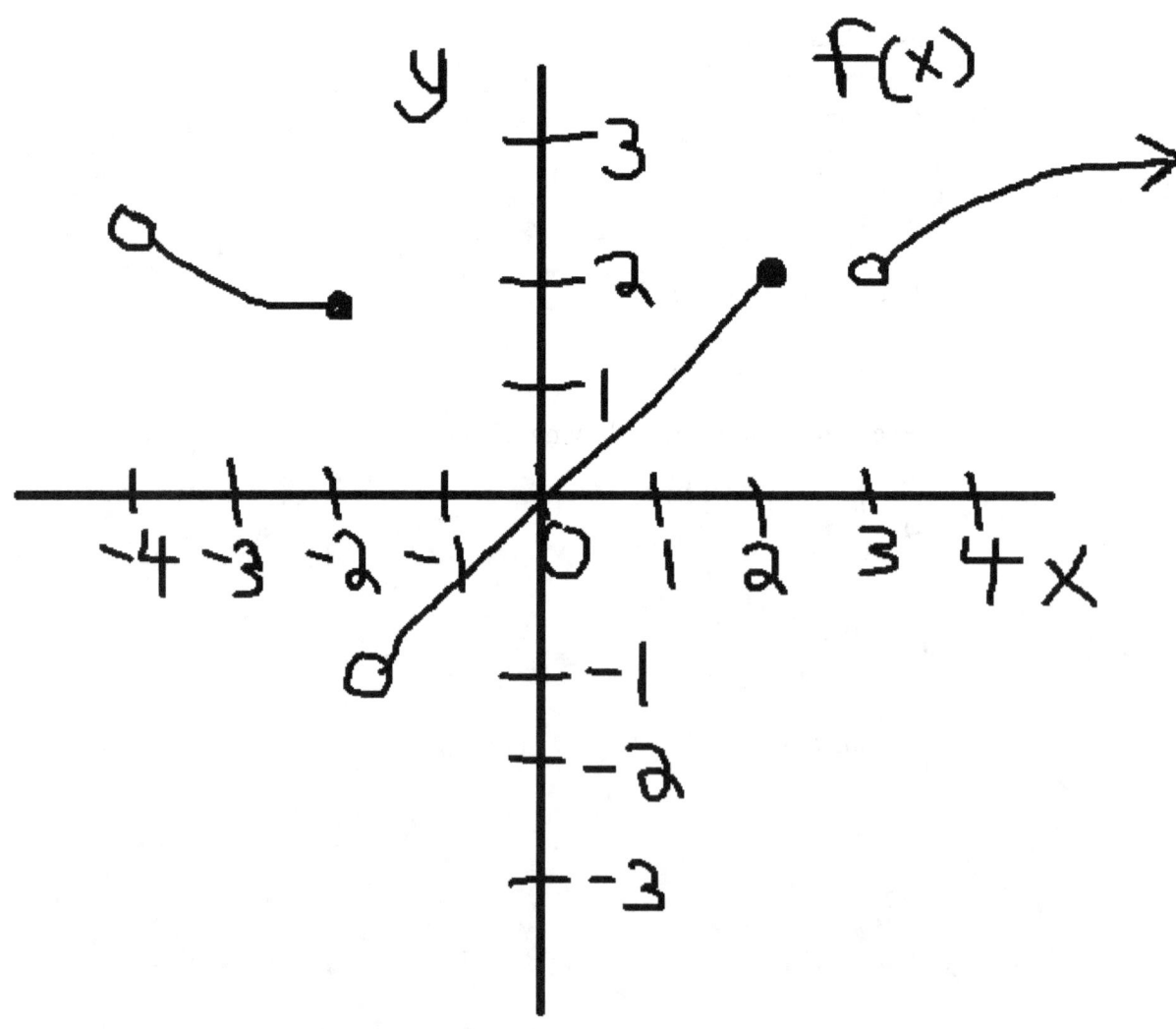

CONCEPTUAL EXERCISES #11

1. True or False: $\lim\limits_{x \to a} f(x) = L$ if and only if $\lim\limits_{x \to a^-} f(x) = \lim\limits_{x \to a^+} f(x) = L$

2. True or False: If $\lim\limits_{x \to a} f(x) = f(a)$ and $f(a)$ is defined, then $f(x)$ is continuous at $x = a$.

3. True or False: If $f(a)$ is undefined then $\lim\limits_{x \to a} f(x)$ does not exist.

4. True or False: Horizontal asymptotes can only occur at $x = a$ if $\lim\limits_{x \to a} f(x) = \pm\infty$.

5. True or False: $\lim\limits_{x \to a} \sqrt[n]{f(x)g(x)} = \sqrt[n]{\lim\limits_{x \to a} f(x)} \sqrt[n]{\lim\limits_{x \to a} g(x)}$.

6. True or False: If $f(x) = p(x) / q(x)$, and the degree of $p(x)$ equals the degree of $q(x)$, then there is a horizontal asymptote at $y = 0$.

7. True or False: The Intermediate Value Theorem can be applied to find a value c in $[a, b]$ so that $f(c) = 0$ if $f(x)$ is continuous in $[a, b]$ and $f(a) < f(b)$ or $f(a) > f(b)$.

8. True or False: If $f(x) \geq g(x)$ for all x, then $\lim\limits_{x \to a} f(x) \geq \lim\limits_{x \to a} g(x)$.

9. True or False: The three types of discontinuities are jump, point, and asymptotic. Point discontinuities can also be removable.

10. True or False: The Squeeze/Sandwich Theorem can be used to show that $\lim\limits_{x \to \infty} \dfrac{\cos x}{x} = 0$.

11. Prove that $\lim\limits_{x \to 5} 2x - 4 = 6$.

12. Prove that $\lim\limits_{x \to 4} x^2 - 9 = 7$.

13. Use the Squeeze/Sandwich Theorem and the diagram below to show that $\lim\limits_{\theta \to 0} \dfrac{\sin\theta}{\theta} = 1$. Hint: The area of a sector $= \dfrac{1}{2}\theta r^2$.

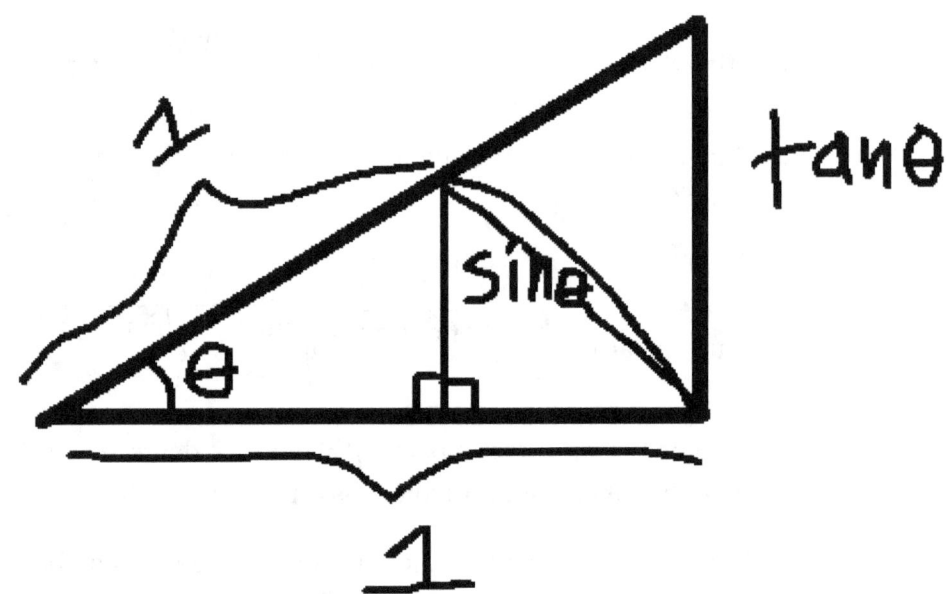

14. Show that $\lim\limits_{\theta \to 0} \dfrac{1 - \cos\theta}{\theta} = 0$.

15. Prove each of the following limit properties given that $\lim\limits_{x \to a} f(x) = L$ and $\lim\limits_{x \to a} g(x) = M$.

a) $\lim\limits_{x \to a}[f(x) + g(x)] = L + M$

b) $\lim\limits_{x \to a}[f(x)]^n = L^n$

156

16. Verify that the statement is true. If $\lim\limits_{x \to a} f(x) = \infty$, then $\lim\limits_{x \to a} \dfrac{1}{f(x)} = 0$.

17. Show that $\lim\limits_{x \to \infty} \dfrac{c}{x} = 0$, for any real number c.

18. Use the Intermediate Value Theorem to show that the polynomial function $f(x) = x^3 + x^2 - 3x + 2$ has a zero in $[-4, 0]$.

19. Let $f(x) = x^2 - 9$ on the interval $[2, 5]$. Show there exists a value of c between 2 and 5 so that $f(c) = 6$.

20. Suppose that $f(x)$ and $g(x)$ are continuous functions on $[a, b]$. If $f(a) < g(a)$ and $f(b) > g(b)$, is there a value c in (a, b) such that $f(c) = g(c)$? Explain your reasoning and give an example to support your claim.

REVIEW #12

CALCULUS 1 REVIEW PART 2

The first part of the calculus 1 review defines what a limit is, how to compute limits, the relationship between limits and continuity of functions, and some important consequences of these ideas. In part 2 of the calculus 1 review, we are going take the concept of a limit a step further to find a solution to the slope of the tangent line problem or the derivative. We will also review derivative rules, implicit differentiation, and the derivatives of trigonometric functions and their inverses as well as exponential and logarithmic functions.

TANGENT LINE PROBLEM

One of the questions that puzzled mathematicians for many years was how to find the slope of a tangent at a given point on a curve. Finding the slope of a line given two points was fairly easy with the help of algebra. But if we apply the concept of the limit to the algebraic formula of slope, then we can definitely solve this problem.

Suppose we have an arbitrary function (non-linear) $f(x)$ and we define two points on $f(x)$, which are $(a, f(a))$ and $(x, f(x))$. If we connect these two points together, we create a secant line. We use the slope formula to calculate the slope m of the secant line.

$$m = \frac{f(x) - f(a)}{x - a}$$

Now suppose that x is very close to a. If we let $x = a + h$, where h is a very small number close to zero, then we have the following limit for computing the slope at $x = a$ or m_a.

$$m_a = \lim_{h \to 0} \frac{f(a + h) - f(a)}{h}$$

This is the slope of the tangent line at $x = a$. If we want to find the slope of the tangent line of $f(x)$ for any given value of x, then we can define this as

$$f'(x) = \lim_{h \to 0} \frac{f(x + h) - f(x)}{h} .$$

We call $f'(x)$ the derivative of $f(x)$ with respect to x. The notation $f'(x)$ is read as "f prime of x." In order to find the derivative of any function, we use this definition. The process of finding the derivative is called "differentiation."

A derivative can also be thought of as an instantaneous rate of change at a given point on the curve. We may want to know how fast or slow the rate of change of something is at that instant.

Now functions may or may not have derivatives at certain points on their graphs. This brings up the question of differentiability, which is our next definition.

DIFFERENTIABILITY: A function $f(x)$ is differentiable at $x = a$, if $f'(x)$ exists and $f(x)$ is differentiable on an interval if $f'(x)$ exists for all points inside the interval.

We can relate differentiability and continuity of functions with the following theorem.

THEOREM OF DIFFERENTIABILITY/CONTINUITY: If $f(x)$ is differentiable at $x = a$, then $f(x)$ is continuous at $x = a$.

Here is a proof of why this is true.

PROOF: Let $f(x)$ be differentiable at $x = a$.

Then $f'(a) = \lim_{x \to a} \dfrac{f(x) - f(a)}{x - a}$.

Now if $x \neq a$, we have $f(x) - f(a) = \dfrac{f(x) - f(a)}{x - a}(x - a)$.

Then $\lim_{x \to a} f(x) - f(a) = \lim_{x \to a} \dfrac{f(x) - f(a)}{x - a} \lim_{x \to a}(x - a) = f'(a) \cdot 0 = 0$.

So, $\lim_{x \to a} f(x) - f(a) = 0$.

Now $f(x) = f(x) - f(a) + f(a)$.

Then $\lim_{x \to a} f(x) = \lim_{x \to a} f(x) - f(a) + f(a) = 0 + f(a) = f(a)$.

Therefore, $f(x)$ is continuous at $x = a$. *

What about the converse of this statement? Is it true that if a function is continuous at $x = a$, then it is differentiable at $x = a$? The student will answer this question as an exercise at the end of this review.

Let us look at some important derivative rules that follow from the definition of the derivative. These rules will help us easily find the derivative of functions without resorting to the definition all the time, and grinding through several, tedious steps of algebra.

DERIVATIVE RULES

SUM/DIFFERENCE RULE: $\dfrac{d}{dx}[f(x) \pm g(x)] = f'(x) \pm g'(x)$

CONSTANT TIMES A FUNCTION RULE: For any constant c,
$\dfrac{d}{dx}[cf(x)] = c\,f'(x)$.

CONSTANT RULE: For any constant c, $\dfrac{d}{dx}[c] = 0$.

POWER RULE: For any real number n, $\dfrac{d}{dx}[x^n] = nx^{n-1}$

PRODUCT RULE: $\dfrac{d}{dx}[f(x)g(x)] = f(x)g'(x) + f'(x)g(x)$.

QUOTIENT RULE: $\dfrac{d}{dx}\left[\dfrac{f(x)}{g(x)}\right] = \dfrac{f'(x)g(x) - f(x)g'(x)}{[g(x)]^2}$.

CHAIN RULE: If $f(x)$ and $g(x)$ are both differentiable functions, then
$\dfrac{d}{dx}[f(g(x))] = f'(g(x))g'(x)$

It is important to note that $f'(x)$ can also be written $\dfrac{df}{dx}$, which is known as the Leibniz notation named after German mathematician Gottfried Leibniz who was one of the inventors of calculus.

One a side note, English mathematician Isaac Newton also invented calculus independently of Leibniz. But Leibniz was the first to publish the results of calculus. However, both mathematicians fought tooth and nail to receive credit for their discoveries, which they eventually did.

All of the proofs of the derivative rules except the chain rule will be left as exercises for the student. We will give an informal, less rigorous proof of why the chain rule is true using the Leibniz notation.

Let $y = f(u)$, where $u = g(x)$ and suppose $f(u)$ and $g(x)$ are both differentiable. Then $\dfrac{dy}{dx} = \dfrac{dy}{du}\dfrac{du}{dx}$.

PROOF: Let $\Delta y = f(u + \Delta u) - f(u)$ and $\Delta u = g(x + \Delta x) - g(x)$. Then we can write $\dfrac{\Delta y}{\Delta x} = \dfrac{\Delta y}{\Delta u}\dfrac{\Delta u}{\Delta x}$. Taking the limit of both sides, we have

$$\lim_{\Delta x \to 0}\frac{\Delta y}{\Delta x} = \lim_{\Delta x \to 0}\frac{\Delta y}{\Delta u}\lim_{\Delta x \to 0}\frac{\Delta u}{\Delta x}.$$ Since $g(x)$ is differentiable, it is also continuous.

So, we have $\Delta u \to 0$ as $\Delta x \to 0$.

It follows that $$\lim_{\Delta x \to 0}\frac{\Delta y}{\Delta x} = \lim_{\Delta u \to 0}\frac{\Delta y}{\Delta u}\lim_{\Delta x \to 0}\frac{\Delta u}{\Delta x}.$$

Thus, we have $\dfrac{dy}{dx} = \dfrac{dy}{du}\dfrac{du}{dx}$. *

An important consequence of the chain rule is another form of differentiation we can use to solve for the derivative. This form of differentiation is called "implicit differentiation." Implicit differentiation can be used to find the derivative of equations that have both the variables x and y together. In other words, the variable y is not solved explicitly in terms of x. Here are some basic rules of implicit differentiation.

VARIABLE RULE: $\dfrac{d}{dx}[y] = \dfrac{dy}{dx}$

POWER RULE: For any real number n, $\dfrac{d}{dx}[y^n] = ny^{n-1}\dfrac{dy}{dx}$

Implicit differentiation can be used for the other rules such as product rule, quotient rule, etc.

How do we solve equations like $x^2 + xy = 1$ with both variables x and y? Here is a simple 3 step-by-step approach to solving such equations.

SOLVING EQUATIONS BY IMPLICIT DIFFERENTATION

1. Take the derivative of both sides respect to x. (Take the derivative of each term on both sides).

2. Get the terms with $\frac{dy}{dx}$ on the same side. Treat it like an unknown variable in solving regular equations in algebra.

3. Cancel out like terms, factor if necessary and solve for $\frac{dy}{dx}$.

When we do all three steps very carefully, we can easily find the derivative of our function.

It is important to note that since $y = f(x)$, we can write the derivative of $f(x)$ as y'.

DERIVATIVES OF TRIGONOMETRIC AND INVERSE TRIGONOMETRIC FUNCTIONS

We are going to list the derivatives of the six trigonometric functions. The details to why these are true are left to the student as exercises.

$$\frac{d}{dx}(\sin x) = \cos x$$

$$\frac{d}{dx}(\cos x) = -\sin x$$

$$\frac{d}{dx}(\tan x) = \sec^2 x$$

$$\frac{d}{dx}(\csc x) = -\csc x \cot x$$

$$\frac{d}{dx}(\sec x) = \sec x \tan x$$

$$\frac{d}{dx}(\cot x) = -\csc^2 x$$

$$\frac{d}{dx}(\sin^{-1} x) = \frac{1}{\sqrt{1-x^2}}$$

$$\frac{d}{dx}(\cos^{-1} x) = -\frac{1}{\sqrt{1-x^2}}$$

$$\frac{d}{dx}(\tan^{-1} x) = \frac{1}{1+x^2}$$

$$\frac{d}{dx}(\csc^{-1} x) = -\frac{1}{|x|\sqrt{1-x^2}}$$

$$\frac{d}{dx}(\sec^{-1} x) = \frac{1}{|x|\sqrt{1-x^2}}$$

$$\frac{d}{dx}(\cot^{-1} x) = -\frac{1}{1+x^2}$$

EXPONENTIAL AND LOGARITHMIC DERIVATIVES

We can apply the definition of the derivative and/or the chain rule to determine the derivatives of exponential and logarithmic functions. One of the most famous and beloved functions (in my humble opinion) of calculus is $f(x) = e^x$. We will see why this may be true to many members of the mathematical and scientific communities.

$$\frac{d}{dx}(e^x) = e^x$$

Wow! The derivative of e^x is e^x. This result is interesting, but why? Here is a proof using the definition of the derivative.

PROOF:

$$\frac{d}{dx}(e^x) = \lim_{h \to 0} \frac{e^{x+h} - e^x}{h} = \lim_{h \to 0} \frac{e^x(e^h - 1)}{h}$$

Now as $h \to 0$, $e^h \to 1 + h$. So, we rewrite this as

$$\lim_{h \to 0} \frac{e^x(1 + h - 1)}{h} = \lim_{h \to 0} \frac{e^x h}{h} = e^x. \ *$$

Note that the h's cancel since $h \neq 0$.

Now we move on to finding the derivative of $f(x) = \ln x$.

$$\frac{d}{dx}(\ln x) = \frac{1}{x}$$

Why is this true? We can provide a quick proof using implicit differentiation. The proof by definition will be left as an exercise.

PROOF:

If $y = \ln x$, then $e^y = e^{\ln x} = x$. Then $\frac{d}{dx}(e^y) = \frac{d}{dx}(x)$ implies that

$e^y \frac{dy}{dx} = 1$ or $\frac{dy}{dx} = \frac{1}{e^y}$. But $e^y = x$, so $\frac{dy}{dx} = \frac{1}{x}$. *

Implicit differentiation is a useful tool for finding the derivatives of exponential functions and inverse trigonometric functions.

Two other functions where implicit differentiation can be used to the find the derivative is $f(x) = a^x$ and $f(x) = \log_a x$. The proofs will be left as exercises.

$$\frac{d}{dx}(a^x) = a^x \ln a$$

$$\frac{d}{dx}(\log_a x) = \frac{1}{x \ln a}$$

Another tactic that we employ when finding the derivative of exponential functions like $f(x) = x^x$ is logarithmic differentiation. This process involves taking the natural log of both sides, differentiating both sides, and solving for y'. We can see that logarithmic differentiation is an extended version of implicit differentiation.

Other functions that are useful in calculus, engineering, and physics are the hyperbolic functions and their inverses. The derivatives of these functions are omitted from the text, but the student can prove a subset of the hyperbolic functions as exercises.

This wraps up the second part of calculus 1 on the definition and rules of the derivative. The third part will cover more topics on the derivative such as higher ordered derivatives, some applications of how to use the derivative to determine the behavior and shape of graphs, and some important theorems that build the foundation for the application of the derivative.

COMPUTATIONAL EXERCISES #12

1. Use the definition of the derivative to find the derivative of each of the following functions. Also, evaluate $f'(0), f'(2), f'(-1), f'(-5)$.

   a) $f(x) = 2x + 3$

   b) $f(x) = x^2 - 4x + 1$

   c) $f(x) = \sqrt{x}$

   d) $f(x) = \dfrac{1}{x+5}$

2. Find the derivative of $f(x) = |x|$. Is $f(x)$ differentiable at $x = 0$?

3. Use the power rule and constant rule to find the derivative of each function.

   a) $f(x) = 6x^4 - 8x^3 + x - 10$

   b) $f(x) = \sqrt{x^3} + 7\sqrt[3]{x^2}$

4. Use the product rule to find the derivative of each function. Simplify if necessary.

   a) $f(x) = x^2 \ln x$

   b) $f(x) = e^x(\sin x - \cos x)$

5. Use the quotient rule to find the derivative of each function. Simplify if necessary.

   a) $f(x) = \dfrac{4x}{x-2}$

   b) $f(x) = \dfrac{\sec x}{\csc x + \cot x}$

6. Use the chain rule to find the derivative of each function.

a) $f(x) = (x^2 + 3x)^5 - \dfrac{1}{8 - x}$

b) $f(x) = \sin(2x) + e^{4x-2} - \ln(x^3)$

7. Use a combination of any of the differentiation rules to find the derivative of each function. Simplify if necessary.

   a) $f(x) = \dfrac{(x^2 + x)(x + 1)}{x - 1}$

   b) $f(x) = e^{\sin x}(\cos x + \sin x)$

   c) $f(x) = \log_2(7x - 14) - \log_2(x^2 - 2x)$

   d) $f(x) = 3^{\ln(6x)}$

   e) $f(x) = \ln(\tan x) + \ln(\sec x)$

   f) $f(x) = \dfrac{e^x + e^{-x}}{e^x - e^{-x}}$

8. Solve for $\dfrac{dy}{dx}$ or y' in each of the following equations.

   a) $xy + 2x = 1$

   b) $(x - y)^2 = y + 3$

   c) $4x^2y + 8x = 10y - xy^2$

   d) $x + y = x^y$

CONCEPTUAL EXERCISES #12

1. True or False: A derivative can be thought of as the slope of the tangent line at a given point on the curve or the instantaneous rate of change at a given instant.

2. If $f(x)$ is a continuous function, then $f(x)$ is also differentiable.

3. True or False: The derivative of the sum of functions $f(x)$ and $g(x)$ is the sum of their individual derivatives.

4. True or False: The derivative of constant times a function is always zero.

5. True or False: If c is any real number and $f(x) = c$, then $f'(x) = 0$.

6. True or False: The function $f(x) = x^{1/n}$ is differentiable for any natural number n.

7. True or False: The derivative of an even function is an odd function and vice versa.

8. True or False: If $g(a) = c$ and $g'(a) = 0$, then $[f(g(x))]' = 0$.

9. True or False: A function can be differentiable at points where there are sharp points, cusps, and vertical tangents.

10. True or False: Another way to express the derivative of $y = f(x)$ is y' or the Leibniz notation $\dfrac{dy}{dx}$.

11. Prove all of the derivative rules except the chain rule using the definition of the derivative.

12. Prove the derivatives of $\sin x$ and $\cos x$ by definition of the derivative.

13. Prove the derivatives of the rest of the trigonometric functions using the derivative rules.

14. Using the definition of the derivative to prove the derivative of $\ln x$.

15. Use logarithmic differentiation to prove the derivatives of a^x and $\log_a x$.

16. If $x^y = y$, then find y'. At what point(s) is $y' = 0$? Undefined?

17. If $g(x) = e^x f(x)$ and $f(c) = -f'(c)$, then what is $g'(c)$?

18. The hyperbolic functions $\sinh x = \dfrac{e^x - e^{-x}}{2}$ and $\cosh x = \dfrac{e^x + e^{-x}}{2}$ are utilized in many areas of mathematics and science.

 a) Show that the derivative of one hyperbolic function is the other hyperbolic function.

 b) Show that $\cosh^2 x - \sinh^2 x = 1$.

 c) Using the result from b), show that $\dfrac{d}{dx}(\cosh^{-1} x) = \dfrac{1}{\sqrt{x^2 - 1}}$ and

 $\dfrac{d}{dx}\left(\sinh^{-1} x\right) = \dfrac{1}{\sqrt{x^2 + 1}}$. Are any of these derivatives related to the

 derivatives of the inverse trigonometric functions?

19. Prove or disprove. $\dfrac{d}{dx}(\cos^{-1}(\sin x)) = \dfrac{d}{dx}(\sin^{-1}(\cos x))$

20. Show that $\dfrac{d}{dx}(\cos^{-1} x) + \dfrac{d}{dx}(\sin^{-1} x) = 0$.

 Verify that $\cos^{-1} x + \sin^{-1} x = \dfrac{\pi}{2}$.

REVIEW #13

CALCULUS 1 PART 3

The third part of the calculus 1 review will go over higher derivatives and their various notations as well as applications of the derivative that includes some very useful definitions and theorems to lay the foundation. We will learn how the first and second derivatives affect the shape of a graph for a function.

In part 2 of the calculus 1 review, we learned the definition of the definition and how to compute derivatives by using various rules. In calculus, we can also calculate higher ordered derivatives, i.e. second derivative, third derivative, all the way up to the nth derivative. Below are some ways we can express higher ordered derivatives.

1st Derivative: $f'(x), y', \dfrac{dy}{dx}$

2nd Derivative: $f''(x), y'', \dfrac{d^2 y}{dx^2}$

3rd Derivative: $f'''(x), y''', \dfrac{d^3 y}{dx^3}$

4th Derivative: $f^{(4)}(x), y^{(4)}, \dfrac{d^4 y}{dx^4}$

nth Derivative: $f^{(n)}(x), y^{(n)}, \dfrac{d^n y}{dx^n}$

Notice that once we get past the third derivative, we do not use the prime notation for $f(x)$ and y, but write the nth derivative in parentheses.

LINEAR APPROXIMATION

There are some functions $f(x)$ that we may have difficulty evaluating for certain values of x. We can use the derivative to find the equation of the tangent line at a given point $x = a$. Once we have the equation of the tangent line, we can estimate $f(x)$ for values of x near a. Let us derive the equation of the tangent line of $f(x)$ at $x = a$ by applying the slope formula.

$y - y_1 = m(x - x_1)$. Now if we let $m = f'(a)$, $x_1 = a$ and $y_1 = f(a)$, then we have

$y - f(a) = f'(a)(x - a)$. Solving for y gives us

$y = f(a) + f'(a)(x - a)$. We can also express this as

$f(x) \approx f(a) + f'(a)(x - a)$.

This is known as the linearization of the function $f(x)$ at $x = a$.

If we look at the figure below, we can see that the linearization of $f(x)$ can give us a good estimation for values of x near a. As x gets farther away from $x = a$, we do not receive a good estimation for $f(x)$.

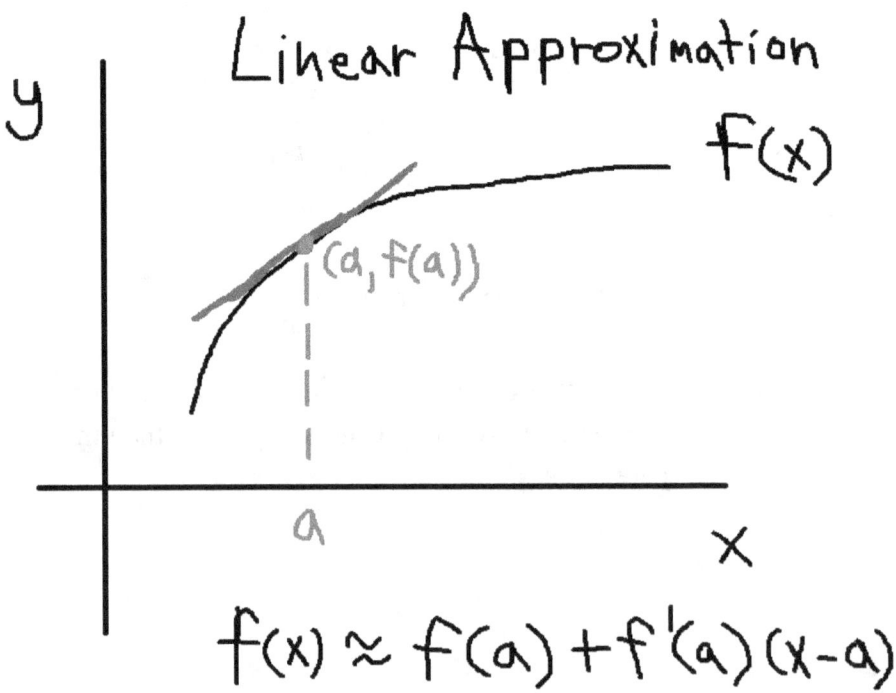

Linear Approximation

$f(x) \approx f(a) + f'(a)(x-a)$

170

DIFFERENTIALS

We know that if $y = f(x)$, then $\dfrac{dy}{dx} = f'(x)$ follows from taking the derivative. Now we can multiply dx to both sides to get $dy = f'(x)dx$. We call both dy and dx differentials. If we know the derivative of $f(x)$ at a certain value of x and the value of dx, then we can the value of dy. This shows us how y is changing in terms of the derivative and the change in x.

Now there is an important application to differentials. We can compute algebraically the change in y in terms of the change in x. This formula gives us $\Delta y = f(x + \Delta x) - f(x)$. However, there are some functions where Δy is difficult to compute. If we let Δx get very small and close to zero, then we see that $dy \approx \Delta y$. We know that dy is very easy to calculate since we have the derivative of $f(x)$ and the value of dx. Observe the figure that illustrates the concept of differentials.

Differentials

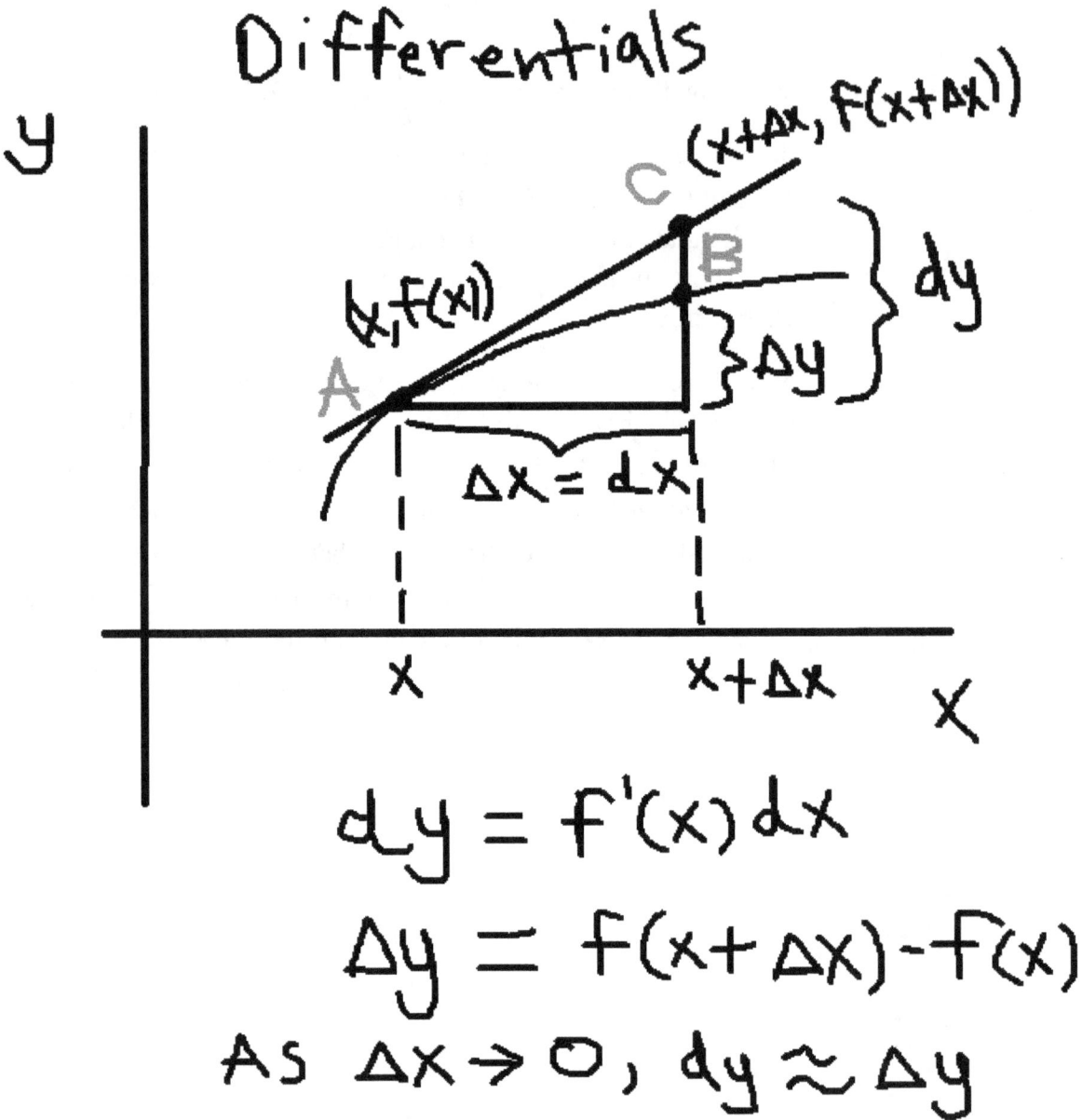

$$dy = f'(x)\,dx$$

$$\Delta y = f(x + \Delta x) - f(x)$$

$$\text{AS } \Delta x \to 0, \; dy \approx \Delta y$$

We can also calculate the linearization of $f(x)$ by using differentials. We can use the formula $f(a + dx) \approx f(a) + dy = f(a) + f'(x)dx$.

Another important application to differentials involves the accuracy and precision of measurements. We can find the maximum error of certain measurements such as distance, area, volume, etc. For instance, if $y = f(x)$ represents the formula of a particular measurement, then $dy = f'(x)dx$ represents the maximum error of that measurement.

172

To find the percent error in comparison of the two measurements, we have $error = \dfrac{dy}{y} \times 100 = \dfrac{f'(x)dx}{f(x)} \times 100$.

PHYSICS APPLICATION TO DERIVATIVES

One important application of higher ordered derivatives is used in physics to describe the motion of an object. Any object has a certain position, velocity, and acceleration at a given instant.

If we let $s(t)$ be the position function of the object at time t, then as the position of the object changes over a period of time, then the object has a certain velocity function $v(t)$ at time t. Velocity is the speed and the direction of the object. And if the object's velocity changes over a period of time, then the object has a certain acceleration function $a(t)$ at time t. Acceleration is how fast the speed of the object is changing over time. Now how does the position, velocity, and acceleration functions relate to each other in terms of derivatives?

If we want to know the instantaneous rate of change of the object's position $s(t)$ at a given time t, then $v(t) = s'(t)$. In other words, the velocity of the object is the first derivative of the object's position.

If we want to know the instantaneous rate of change of the object's velocity $v(t)$ at a given time t, then $a(t) = v'(t) = s''(t)$. In other words, the acceleration of the object is the first derivative of the object's velocity and the second derivative of the object's position.

We need a way of interpreting the signs of position, velocity, and acceleration in one-dimensional motion. Look at the table.

SIGNS	$s(t)$	$v(t)$	$a(t)$
POSITIVE	Object is to the right of the origin.	Object is moving to the right.	Object is speeding up.
NEGATIVE	Object is to the left of the origin.	Object is moving to the left.	Object is slowing down.
ZERO	Object is at the origin.	Object is at rest.	Object is moving at constant speed or is at rest.

SIDE NOTES: Speed and velocity are two different concepts. Speed is how fast or slow an object is traveling. Velocity is speed and the direction in which the object is traveling. In other words, speed is a scalar quantity, and velocity is a vector quantity. Also, an object can be moving at a constant speed, but may or may not accelerate. If the object is moving at a constant speed and changing direction, then the object is accelerating. But if the object is moving at a constant speed and is not changing direction, then the object is not accelerating. In order for the object to accelerate, either the speed and/or direction components of its velocity must change.

We are going to switch gears and discuss some important definitions and theorems in calculus that will help us understand the behavior and shapes of the graphs of continuous functions.

MAXIMUM/MINIMUM VALUES OF A FUNCTION

A very important application in calculus that engineers and scientists use on a daily basis is optimization. Optimization is the process of finding the best way to do something. Companies want their employees and staff to maximize productivity and minimize cost in the workplace. Calculus helps us to find these maximum and minimum values of a function to produce optimal results.

Suppose that we have a continuous function $f(x)$ on $[a, b]$. The function may or may not have some maximum and/or minimum values within this interval or the maximum and/or minimum values may be located at the endpoints a and b. The collection of the maximum and minimum values of $f(x)$ is called the "extrema" or extreme values of $f(x)$. A maximum or minimum value may be absolute or relative, depending on where the extreme values are located on the graph. Below are some definitions of absolute/relative maximum and minimum.

ABSOLUTE MAXIMUM: If $f(c) \geq f(x)$ for all x in the domain of f, then $f(x)$ has an absolute maximum at $x = c$ and $f(c)$ is the absolute maximum.

ABSOLUTE MINIMUM: If $f(c) \leq f(x)$ for all x in the domain of f, then $f(x)$ has an absolute minimum at $x = c$, and $f(c)$ is the absolute minimum.

RELATIVE MAXIMUM: If $f(c) \geq f(x)$ for values of x near c, then $f(x)$ has a relative maximum at $x = c$ and $f(c)$ is the relative maximum.

RELATIVE MINIMUM: If $f(c) \geq f(x)$ for values of x near c, then $f(x)$ has a relative minimum at $x = c$ and $f(c)$ is the relative minimum.

Note that an absolute maximum or minimum can occur for any values on $[a, b]$ including the endpoints. But a relative maximum or minimum can occur only for any values near $x = c$, but cannot occur on the endpoints.

These definitions lead us to an important theorem of extreme values of any continuous function.

EXTREME VALUE THEOREM: If $f(x)$ is continuous on $[a, b]$, then there exists numbers c and d in $[a, b]$ such that $f(c)$ is the absolute maximum and $f(d)$ is the absolute minimum.

This theorem seems very intuitive, but is quite difficult to prove. So, the proof is omitted from the text.

Notice that we can only use the extreme value theorem if $f(x)$ is continuous for all values of x in $[a, b]$. Like the intermediate value theorem, the extreme value theorem is also an existence theorem that tells us that absolute maximum/minimum values exists, but does not tell us where they are located on the graph. Here is a graph that illustrates the extreme value theorem.

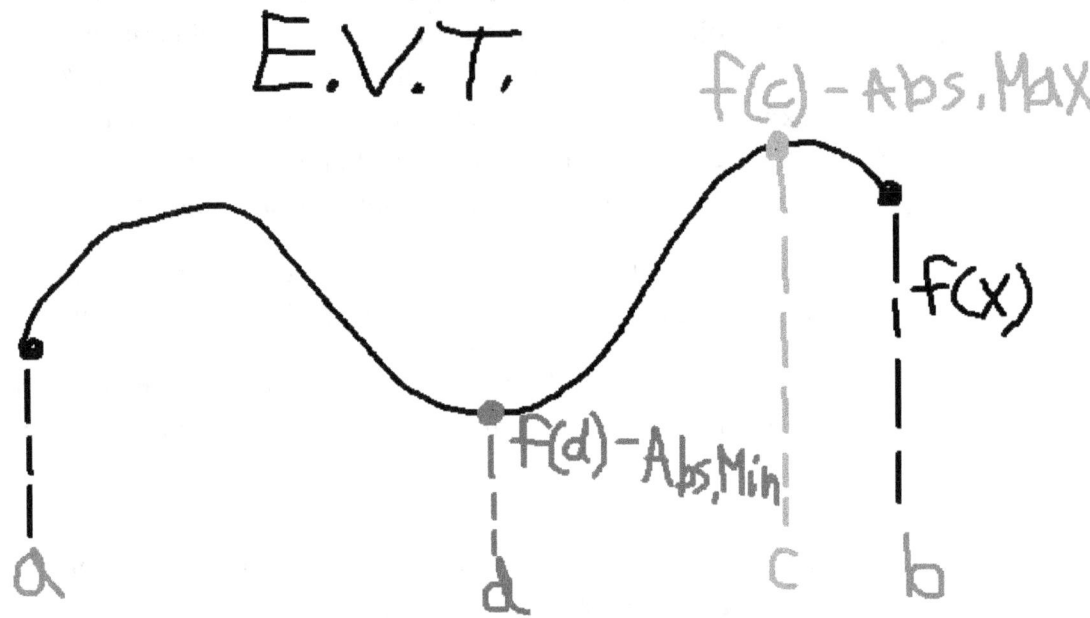

If we look at the graph we see that the absolute maximum $f(c)$ and absolute minimum $f(d)$ both have horizontal tangent lines. And the derivative of a horizontal tangent line is zero. Thus, $f'(c) = f'(d) = 0$. This idea leads us to Fermat's Theorem.

FERMAT'S THEOREM: If $f(x)$ has a relative maximum or minimum at $x = c$ and $f'(c)$ exists, then $f'(c) = 0$.

This theorem is named after the 17th century French lawyer and mathematician Pierre de Fermat. Fermat was a lawyer by profession, but took up mathematics as a hobby. Along with Descartes, Fermat is one of the founding fathers of analytic geometry. He also made important contributions in number theory with Fermat's Little Theorem and Fermat's Last Theorem, which initially was a conjecture until proven over three centuries later by Andrew Wiles in 1994. Historians made a posthumous discovery of an important note that was written by Fermat on the margin of a page in one of his Greek textbooks that said he found a proof for Fermat's Last Theorem. Unfortunately, Fermat's proof is nowhere to be found and remains a mystery ever since.

We will provide a proof for the relative maximum case. The other case for the relative minimum will be left as an exercise.

PROOF: Suppose that $f(x)$ has a relative maximum at $x = c$. Then $f(c)$ is the relative maximum and $f(c) \geq f(x)$ for all x near c. Let $h > 0$ so that $f(c + h) \leq f(c)$. This implies that $f(c + h) - f(c) \leq 0$. Since h is non-zero, we can divide both sides by h to get $\dfrac{f(c+h)-f(c)}{h} \leq 0$. And since $f'(c)$ exists, we can take the right-hand limit of both sides to get

$$f'(c) = \lim_{h \to 0^+} \frac{f(c+h)-f(c)}{h} \leq \lim_{h \to 0^+} 0 = 0.$$

So, $f'(c) \leq 0$. Now we can do the same for $h < 0$, except the inequality will be reversed. Taking the left-hand limit of both sides, we have

$$f'(c) = \lim_{h \to 0^-} \frac{f(c+h)-f(c)}{h} \geq \lim_{h \to 0^-} 0 = 0.$$

So, $f'(c) \geq 0$. Since $f'(c) \leq 0$ and $f'(c) \geq 0$, we can deduce that $f'(c) = 0$ as required. *

The value of $x = c$ is where a maximum or minimum occurs according to Fermat's Theorem. But what if neither a maximum nor a minimum exists at c or $f(x)$ is not differentiable at c? Then we have the following definition:

CRITICAL NUMBER: A number c in the domain of $f(x)$ is called a critical number if $f'(c) = 0$ or $f'(c)$ does not exist.

In other words, if $f(x)$ has a relative maximum or minimum at $x = c$ or $f(x)$ is not differentiable at $x = c$, then c is a critical number of $f(x)$. It is important to note that if $f(x)$ is continuous on $[a, b]$ then the critical numbers can only be found in (a, b). That is, critical numbers do not exist at the endpoints.

FINDING THE ABSOLUTE MAXIMUM AND ABSOLUTE MINIMUM VALUES OF A FUNCTION

There is a three step process for finding the absolute maximum/minimum values for a function $f(x)$ continuous on $[a, b]$.

1. Find all critical numbers of $f(x)$ in (a, b) and find all $f(c)$ values.

2. Evaluate endpoints $f(a)$ and $f(b)$.

3. The largest value from steps 1 and 2 is the absolute maximum. The smallest value from steps 1 and 2 is the absolute minimum.

Now we are going to move on to a couple of important theorems.

ROLLE'S THEOREM: Let $f(x)$ be a continuous function on $[a, b]$ and a differentiable function on (a, b). If $f(a) = f(b)$, then there exists a c in (a, b) so that $f'(c) = 0$.

In order to use Rolle's Theorem, we must have three conditions to be satisfied. That is 1. $f(x)$ be continuous on $[a, b]$, 2. $f(x)$ is differentiable on (a, b) and 3. $f(a) = f(b)$. If any of these three conditions are not satisfied, then Rolle's Theorem cannot be applied.

According to Rolle's Theorem, since $f'(c) = 0$ this means that there exists a maximum or minimum value in (a, b). Similar to previous theorems, Rolle's Theorem is also an existence theorem. We do not know exactly what the value of c is and whether $f(c)$ is a maximum or a minimum, but we do know that such a value exists and that there is a horizontal tangent line at $x = c$. See the illustration of Rolle's Theorem.

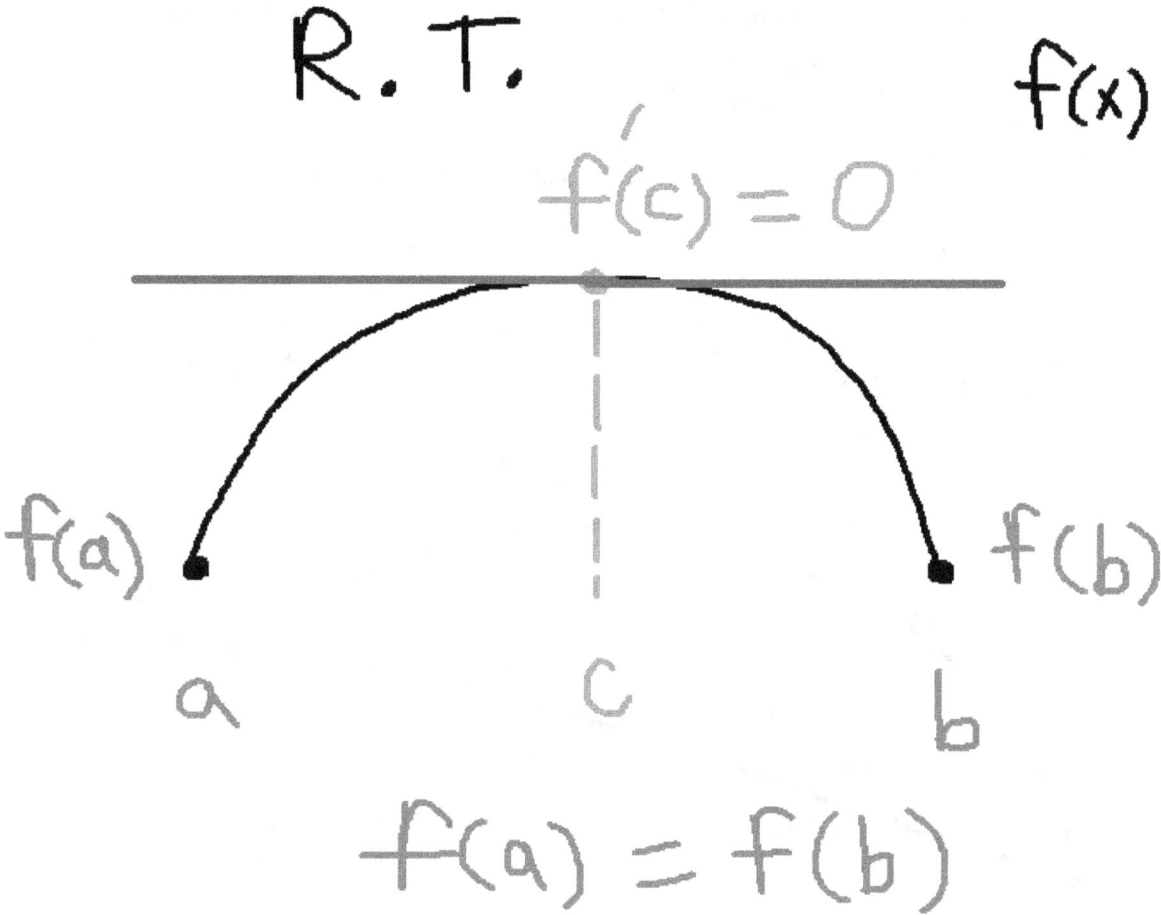

The proof of Rolle's Theorem will be left as an exercise. Now we will look at the general version of Rolle's Theorem or the Mean Value Theorem.

Michel Rolle was a French mathematician in the 17th-18th century who was best known for Rolle's Theorem. What is ironical about Rolle's contributions to calculus is that he was a strong critic of the subject, saying that calculus was a mere collection of ingenious fallacies. However, he later changed his opinion of calculus and accepted the theories to be true. Rolle was also believed to be one of the co-founders of Gaussian Elimination throughout Europe.

MEAN VALUE THEOREM: Suppose $f(x)$ is continuous on $[a, b]$ and differentiable on (a, b). Then there exists a c in (a, b) so that

$$f'(c) = \frac{f(b) - f(a)}{b - a}.$$

Here is an illustration of the Mean Value Theorem.

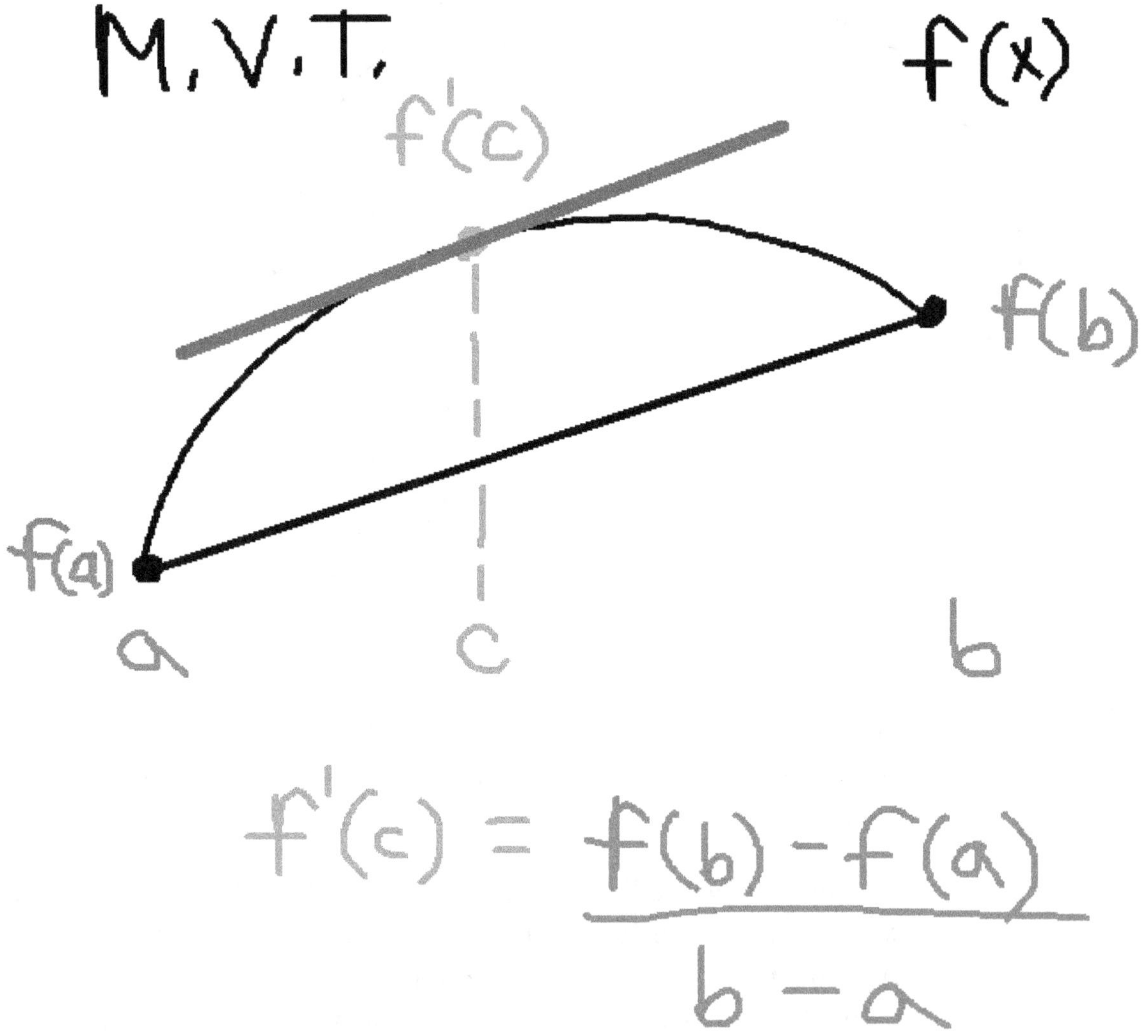

M.V.T.　　　　　　f(x)

f'(c)

f(b)

f(a)

a　　　　c　　　　b

$$f'(c) = \frac{f(b) - f(a)}{b - a}$$

This statement appears to be very intuitive on the surface, but let us understand why this is true.

PROOF: Let $f(x)$ be continuous on $[a, b]$ and differentiable on (a, b). We can define the secant line from a to b as $L_{ab} = f(a) + \dfrac{f(b) - f(a)}{b - a}(x - a)$. Now let us define a new function $g(x) = f(x) - L_{ab}$. Rewriting this, we have

$$g(x) = f(x) - \left[f(a) + \frac{f(b) - f(a)}{b - a}(x - a) \right].$$ Now if we evaluate g(a) and

g(b), we get $g(a) = f(a) - \left[f(a) + \dfrac{f(b) - f(a)}{b - a}(a - a) \right] = 0$ and

$$g(b) = f(b) - \left[f(a) + \frac{f(b) - f(a)}{b - a} \right](b - a) = 0.$$

So, $g(a) = g(b) = 0$. This satisfies the hypothesis of Rolle's Theorem.

Since both $f(x)$ and L_{ab} are both differentiable, then $g(x)$ is also differentiable.

Thus, $g'(x) = f'(x) - \dfrac{f(b) - f(a)}{b - a}$. But by Rolle's Theorem, there exists a value c in (a, b) so that g'(c) = 0. So, $g'(c) = f'(c) - \dfrac{f(b) - f(a)}{b - a} = 0$.

Therefore, $f'(c) = \dfrac{f(b) - f(a)}{b - a}$.

The Mean Value Theorem tells us that the tangent line at $x = c$ between a and b is parallel to the secant line formed between a and b. So, the slopes of both of these lines are equal.

DERIVATIVES AND THE SHAPE OF A GRAPH

HOW TO LOCATE MAXIMUMS AND MINIMUMS

So far, the derivative of a function $f(x)$ has given us some information about the slope of a tangent line at a given point on the curve. But what does $f'(x)$ tell us about $f(x)$? In order to answer this question, let us observe the graph below and see if there is a relationship between the value of $f'(x)$ and the behavior of $f(x)$.

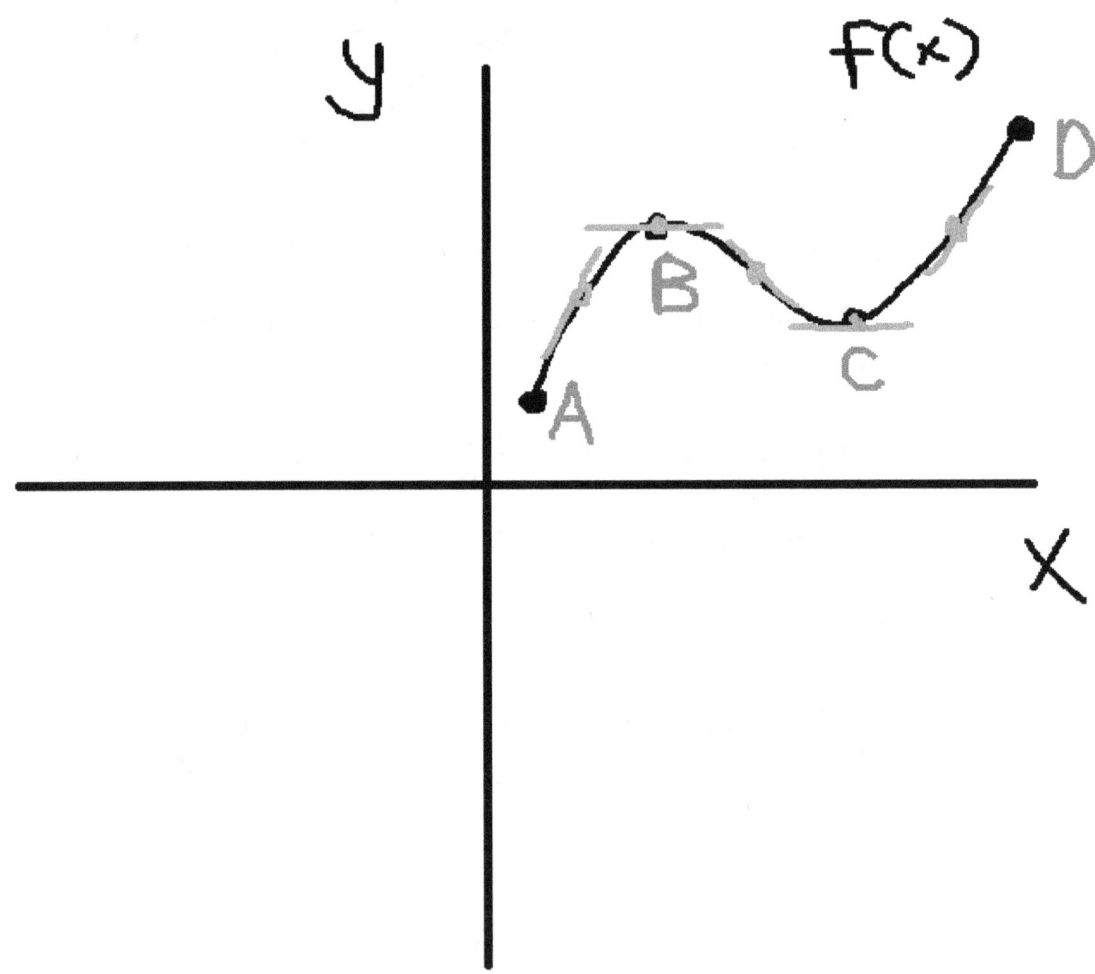

We can describe how the derivative is changing between points. Between points A and B, we see that $f'(x) > 0$ and f(x) is increasing. At point B (B is also a critical point), there is a horizontal tangent, so $f'(x) = 0$. Between points B and C, we know that $f'(x) < 0$ and $f(x)$ is decreasing. Like at point B, we have $f'(x) = 0$ at point C (C is also a critical point). Between points C and D, we see that $f'(x) > 0$ and $f(x)$ is increasing. This leads us to an important test that relates $f'(x)$ to the behavior of $f(x)$.

INCREASING/DECREASING TEST:

1. If $f'(x) > 0$ on any interval I, then $f(x)$ is increasing on I.

2. If $f'(x) < 0$ on any interval I, then $f(x)$ is decreasing on I.

We will provide a proof of the first part of the increasing/decreasing test and leave the details of the second part as an exercise.

PROOF: Let x_1 and x_2 be two numbers so that $x_1 < x_2$. Since $f(x)$ is continuous on $[x_1, x_2]$ and differentiable on (x_1, x_2), then by the Mean Value Theorem, there exists a c in (x_1, x_2) such that

$$f(x_2) - f(x_1) = f'(c)(x_2 - x_1).$$

Now if $x_2 > x_1$, then we have $x_2 - x_1 > 0$. And $f'(c) > 0$ since $f'(x) > 0$ for all x in (x_1, x_2). It follows that $f(x_2) - f(x_1) > 0$ or $f(x_2) > f(x_1)$.

Therefore, $f(x)$ is increasing on $[x_1, x_2]$. *

One thing we notice from the graph above is that when $f'(x) > 0$ between A and B and $f'(x) < 0$ between B and C, we have a maximum at B. Now when $f'(x) < 0$ between B and C and $f'(x) > 0$ between C and D, we have a minimum at C. We can tell whether a critical point will have a maximum or minimum value. This leads us to an important test we can use for the first derivative.

FIRST DERIVATIVE TEST: Suppose $f(x)$ is continuous on $[a, b]$ and differentiable on (a, b) and let $x = c$ be a critical number in (a, b).

1. If $f'(x) > 0$ for all $x < c$ and $f'(x) < 0$ for all $x > c$, then there exists a relative maximum at $x = c$.

2. If $f'(x) < 0$ for all $x < c$ and $f'(x) > 0$ for all $x > c$, then there exists a relative minimum at $x = c$.

3. If $f'(x) > 0$ for all x except $x = c$ or $f'(x) < 0$ for all x except $x = c$, then neither a relative maximum nor relative minimum exists at $x = c$.

In other words, the only way we can be guaranteed relative extrema if the sign of the derivative changes around $x = c$. If the derivative goes from positive to negative around c, then we have a maximum. If the derivative goes from negative to positive around c, then we have a minimum. If the derivative stays positive or stays negative around c, then no maximum or minimum exists.

The Mean Value Theorem establishes some basic facts about how $f'(x)$ is related to $f(x)$.

THEOREM: If $f'(x) = 0$ for all x in (a, b), then $f(x)$ is constant on (a, b).

PROOF: Let x_1 and x_2 be two numbers in (a, b) so that $x_1 < x_2$. Since $f(x)$ is differentiable on (a, b), then it is also differentiable on (x_1, x_2). This implies that $f(x)$ is continuous on $[x_1, x_2]$. Using the Mean Value Theorem, we have a value c in (x_1, x_2) so that

$f(x_2) - f(x_1) = f'(c)(x_2 - x_1)$. Since $f'(x) = 0$ for all x in (a, b), it follows that $f'(c) = 0$. So, $f(x_2) - f(x_1) = 0$ or $f(x_2) = f(x_1)$. Therefore, $f(x)$ is constant on (a, b). *

All this theorem is saying is that since $f'(x) = 0$, we know that $f(x)$ is a constant function in the form $f(x) = c$, where c is a real constant. And any value of x that we plug into the constant function will give us that constant value of c.

Now there is another statement we can use from this theorem to show that the difference between two functions $f(x)$ and $g(x)$ is a constant.

COROLLARY: If $f'(x) = g'(x)$ for all x in (a, b), then $f(x) - g(x)$ is constant on (a, b). In other words, $f(x) = g(x) + C$, for some real constant C.

PROOF: Let $h(x) = f(x) - g(x)$. Then $h'(x) = f'(x) - g'(x) = 0$ since $f'(x) = g'(x)$. Since $h'(x) = 0$, we know that $h(x) = C$, where C is a real constant. Thus, $f(x) - g(x) = C$ or $f(x) = g(x) + C$. *

DERIVATIVES AND THE SHAPE OF A GRAPH

CONCAVITY AND INFLECTION POINTS

We have reviewed what $f'(x)$ says about the original function $f(x)$. Now it is time to review what $f''(x)$ says about $f(x)$. The second derivative tells us about an important aspect of the graph of $f(x)$ known as concavity. For certain intervals, $f(x)$ can either be concave up or concave down. Let us define what concavity means for $f(x)$.

CONCAVITY:

1. If all of the tangent lines lie below $f(x)$ for some interval I, then $f(x)$ is concave up on I.

2. If all of the tangent lines lie above $f(x)$ for some interval I, then $f(x)$ is concave down on I.

Look at the illustration below to get a better understanding of concavity. Notice from point A to point B that $f(x)$ is concave up. From a little past point B to point C, $f(x)$ is concave down. And from a little past point C to point D, $f(x)$ is concave up again. The point where the graph changes concavity from either concave up to concave down or vice versa is called an inflection point. Here is the definition.

CONCAVITY

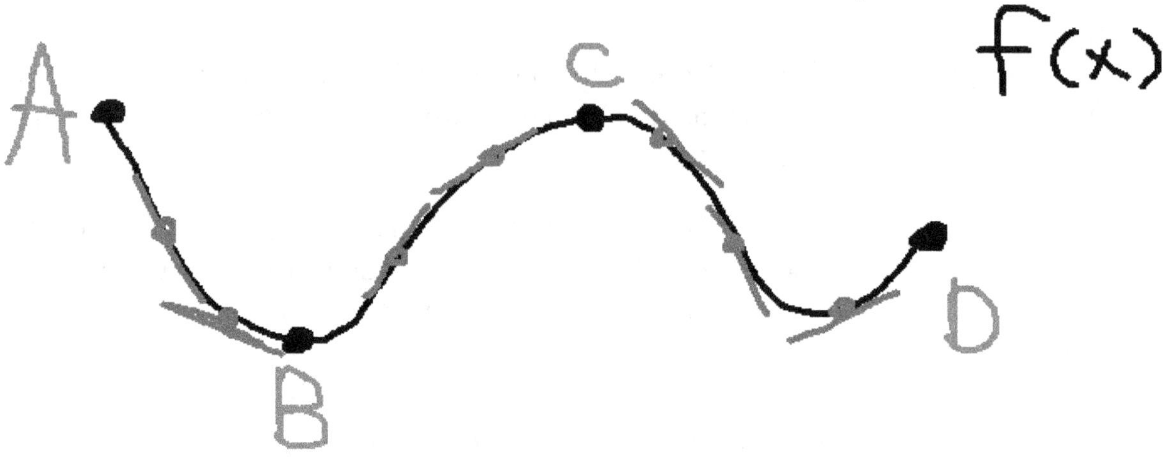

$f(x)$

INFLECTION POINT: If $f''(c) = 0$, then $(c, f(c))$ is an inflection point.

This tells us that if we take the second derivative of the function and find a value c so that $f''(c) = 0$, then we know that the concavity of $f(x)$ begins to change at $(c, f(c))$. Can you spot any points of inflection from the graph above?

But how do we know from the second derivative whether $f(x)$ is concave up or concave down? This question leads us to the concavity test.

CONCAVITY TEST:

1. If $f''(x) > 0$ on an interval I, then $f(x)$ is concave up on I.

2. If $f''(x) < 0$ on an interval I, then $f(x)$ is concave down on I.

We will prove the first case for concave up. The proof of the second case for concave down is similar to the concave up case.

PROOF:

a) Let $x > a$. Since $f(x)$ is differentiable on (a, x), then by the Mean Value Theorem there exists a c in (a, x) such that

$f(x) - f(a) = f'(c)(x - a)$. Since $f''(x) > 0$, this implies that $f'(x)$ is increasing. So, $f'(c) > f'(a)$. Now $x - a > 0$, so we have

$f(a) + f'(c)(x - a) > f'(a)(x - a) + f(a)$ or

$f(a) + f(x) - f(a) > f'(a)(x - a) + f(a)$ or

$f(x) > f'(a)(x - a) + f(a)$. This shows that $f(x)$ lies above all of its tangent lines for $x > a$. So, $f(x)$ is concave up.

b) Let $x < a$. Since $f(x)$ is differentiable on (x, a), then by the Mean Value Theorem there exists a c in (x, a) such that

$f(x) - f(a) = f'(c)(x - a)$. Since $f''(x) > 0$, this implies that $f'(x)$ is increasing. So, $f'(c) < f'(a)$. Now $a - x > 0$, so we have

$f(a) - f'(c)(a - x) > -f'(a)(a - x) + f(a)$ or

(inequality sign reverses)

$f(a) + f'(c)(x - a) > f'(a)(x - a) + f(a)$ or

$f(a) + f(x) - f(a) > f'(a)(x - a) + f(a)$ or

$f(x) > f'(a)(x - a) + f(a)$. This shows that $f(x)$ lies above all of its tangent lines for $x < a$. So, $f(x)$ is concave up. *

Notice in the graph that there are maximum and minimum values. So, we know that there are critical numbers where the maximum and minimum values are located. To get the critical numbers, we have to use the first derivative test. But is there a relationship between the first derivative and second derivative in terms of critical numbers? This brings us to the second derivative test.

SECOND DERIVATIVE TEST: Suppose $f(x)$ is continuous on $[a, b]$ and twice differentiable on (a, b) and $f''(x)$ is continuous on $[a, b]$. Let $x = c$ be a critical number in (a, b) so that $f'(c) = 0$.

1. If $f''(c) < 0$, then there exists a relative maximum at $x = c$.

2. If $f''(c) > 0$, then there exists a relative minimum at $x = c$.

3. If $f''(c) = 0$, then the test is inconclusive.

In other words, if the second derivative is negative at $x = c$, we have a relative maximum. If the second derivative is positive at $x = c$, we have a relative minimum. If the second derivative is zero at $x = c$, then the test fails. Then we revert back to the first derivative test. So, either a relative maximum or relative minimum can occur at $x = c$.

In this review, we have learned many applications to the derivatives from linear approximations and differentials to finding the behavior of a function with maximums/minimums (increasing/decreasing) and inflection points (concavity).

The next review will cover a couple more important applications of the derivative with evaluating limits by L'Hospital's Rule and finding zeros of non-linear functions with Newton's Method. We will also review how to find a function when we know what $f'(x)$ is.

COMPUTATIONAL EXERCISES #13

1. Find the first three derivatives of each of the following functions.

    a) $f(x) = x^4 - 3x^2 + 6x - 2$

    b) $g(x) = (x^3 + 4)^{1/3}$

    c) $h(x) = x^2 \sin x$

    d) $k(x) = \tan^{-1}(x^5)$

2. Find a formula for $f^{(n)}(x)$.

    a) $f(x) = x^n$

    b) $f(x) = e^{4x}$

    c) $f(x) = 1/3x^2$

    d) $f(x) = \sqrt{x-1}$

3. A particle's motion is described by $s(t) = t^4 - t^2 + 4$, where t is in seconds and $s(t)$ is in meters.

a) Find the velocity function $v(t)$ and the particle's velocity after $t = 2$ seconds.

b) Find the acceleration function $a(t)$ and the particle's acceleration after $t = 3$ seconds.

c) When is the particle speeding up? When is it slowing down?

4. The displacement of a mass attached to a vertical spring is given by $y = A \cos \omega t$.

a) Find y' and y''.

b) Verify that the acceleration of the mass is directly proportional to its displacement. What is the constant of proportionality?

c) At what time is the speed of the mass a maximum?

5. If $g(x)$ is a twice differentiable function, then find $f''(x)$ in terms of $g(x)$, $g'(x)$, and $g''(x)$.

a) $f(x) = g(x^3)$

b) $f(x) = x / g(x)$

6. Find the linearization of $f(x)$ for each of the following functions.

a) $f(x) = x^2 - 2x; a = 3$

b) $f(x) = x \cos x; a = \pi$

c) $f(x) = \ln(x + 5); a = -7$

7. Find the differential of each function.

a) $A(x) = x(1 - x^2)^{1/2}$

b) $V(r) = 4/3\pi r^3$

c) $S(t) = v_o t - 1/2 g t^2$

8. Use differentials/linearization to estimate the following values.

a) $\sqrt{4.02}$

b) $\ln(1.35)$

9. Find the critical number(s) of each function.

   a) $f(x) = 4x^3 - 6x^2$

   b) $f(x) = \dfrac{x+5}{x-3}$

   c) $f(x) = \cos^2 x + 2\sin x$

10. Find the absolute maximum and absolute minimum of each function.

    a) $f(x) = x^2 - 2x + 5$ on [0, 3]

    b) $f(x) = -x + \ln x$ on [-1, 1]

11. Use Rolle's Theorem to find a value c so that $f'(c) = 0$.

    a) $f(x) = x^2 - 12x$ on [4, 8]

    b) $f(x) = x\sqrt{x+7}$ on [-7, 0]

12. Use the Mean Value Theorem to find a value c.

    a) $f(x) = x^4 - 8x + 2$ on [0, 1]

    b) $f(x) = \dfrac{3x}{x-1}$ on [2, 4]

13. Find the intervals of increasing/decreasing, maximum and minimum values, intervals of concavity, and inflection points (if possible) for each function.

    a) $f(x) = x + 2\cos x$

    b) $f(x) = \ln x / x$

14. Let $f(x) = x^3 + cx^2 - x$ and suppose that an inflection point occurs at $x = 1$.

   a) What is the value of c?

   b) What are the coordinates for the inflection point?

   c) What are the maximum and minimum values for $f(x)$?

CONCEPTUAL EXERCISES #13

1. True or False: If $f(x) = e^{2x}$, then $f^{(n)}(x) = 2^n e^{2x}$.

2. True or False: The second derivative of the position function $s(t)$ is the acceleration function $a(t)$.

3. True or False: If an object is moving at a constant speed, then the object is not accelerating.

4. True or False: If $f'(c) = 0$, then $f(x)$ has either a maximum or a minimum exists at $x = c$.

5. True or False: A relative maximum/minimum of a function that is continuous on $[a, b]$ and differentiable on (a, b) can also be an absolute maximum/minimum.

6. True or False: Rolle's Theorem can be applied for $f(x) = \sin^{-1}x$ on $[-1, 1]$.

7. True or False: If f is continuous on $[a, b]$ and differentiable on (a, b) and $f(a) = f(b)$, then there exists a value c in (a, b) such that $f'(c) = 0$.

8. True or False: If $f''(x) > 0$ on an interval I, then $f(x)$ is concave up and increasing on I.

9. True or False: If p represents the number of inflection points of a polynomial function and n is the degree of the polynomial, then $p < n$.

10. True or False: If $f'(c) = 0$ and $f''(c) = 0$, then $f(x)$ has neither a maximum nor a minimum at $x = c$.

11. Use implicit differentiation to show that if $xy = x + y$, then
$$\frac{d^2 y}{dx^2} = \frac{2}{1-x} \frac{dy}{dx}.$$

12. Use linearization to show that $\sin x \approx x$ for very small values of x.

13. Prove or Disprove: Let $f(x)$ are $g(x)$ be continuous and differentiable functions for all x. If both $f(x)$ and $g(x)$ are increasing on an interval I, then $f(x) - g(x)$ are also increasing on I.

14. Prove or Disprove. If $f(x)$ is continuous on $[-a, a]$ and differentiable on $(-a, a)$ and $f(x)$ is even, then there exists a c in $(-a, a)$ such that $f'(c) = 0$.

15. Suppose $f(x)$ and $g(x)$ are continuous on $[a, b]$ and differentiable on (a, b). If $f(a) = -g(a)$ and $f'(x) > -g'(x)$, then prove that there exists a c in (a, b) such that $f(b) > -g(b)$. Hint: Let $h(x) = f(x) + g(x)$ and use the Mean Value Theorem.

16. Prove Rolle's Theorem without the aid of the Mean Value Theorem.

17. Prove the second part of the Increasing/Decreasing Test.

18. Prove the second part of the Concavity Test.

19. Show that the largest rectangle that can be inscribed within a semi-circle is a square.

20. Show that the radius R of a cone of maximum volume that can be inscribed in a sphere of radius r is $R = \frac{2}{3} r \sqrt{2}$.

CALCULUS 1 PART 4

In part 4 of the calculus 1 review, we will return back to topic of limits and learn how to evaluate limits with indeterminate forms with the aid of L'Hospital's Rule. We will also cover Newton's Method, which is a useful formula for estimating the zeros of non-linear equations. Finally, we will wrap up the review on the topic of anti-derivatives where we solve the problem of finding the original function $F(x)$ when given the derivative $f(x)$.

INDETERMINATE FORMS AND L'HOSPITAL'S RULE

Right now, we are interesting in computing

$$\lim_{x \to a} \frac{f(x)}{g(x)}.$$

But we run into an obstacle whenever the $f(a) = 0$ and $g(a) = 0$. So, we get $0 / 0$. This is the first type of an indeterminate form. Another problem is when $f(a) = \pm\infty$ and $g(a) = \pm\infty$. So, we have ∞ / ∞. This is the second type of an indeterminate form. How do we evaluate the limit with the two types of indeterminate forms? There is a special rule that we can use to evaluate limits with indeterminate forms called L'Hospital's Rule.

L'HOSPITAL'S RULE: Let $f(x)$ and $g(x)$ be differentiable and $g(x) \neq 0$ near $x = a$, (except maybe at $x = a$).

Suppose that $\lim\limits_{x \to a} f(x) = \lim\limits_{x \to a} g(x) = 0$ or

$\lim\limits_{x \to a} f(x) = \lim\limits_{x \to a} g(x) = \pm\infty.$

If the right-hand limit exists (or is ∞ or $-\infty$), then

$$\lim_{x \to a} \frac{f(x)}{g(x)} = \lim_{x \to a} \frac{f'(x)}{g'(x)}.$$

In essence, L'Hospital's Rule says that if the limit of the quotient of $f(x)$ and $g(x)$ gives us an indeterminate form, then we can take the limit of the quotient of $f'(x)$ and $g'(x)$.

It is important to note that L'Hospital's Rule can only be used for indeterminate forms like $0/0$ and ∞/∞, and both $f(x)$ and $g(x)$ have to be differentiable functions where $g(x) \neq 0$ near $x = a$. L'Hospital's Rule can also be used for one-sided limits provided that all of the conditions are met.

Now another indeterminate form may arise when we evaluate the limit of a difference of functions $f(x)$ and $g(x)$. We can receive a third type of indeterminate form, which is $\infty - \infty$. The solution to this problem is to rewrite the difference of the two functions as a quotient and then use L'Hospital's Rule when appropriate.

The proof for the general form of L'Hospital's Rule is very challenging, which requires the help of a more general version of the Mean Value Theorem known as Cauchy's Mean Value Theorem and proving the cases for when a is a finite real number and when a goes to infinity. This proof is omitted from the text.

We will give a proof for a special case for the first indeterminate form $0/0$. This provides us a little insight to why L'Hospital's Rule is true.

PROOF:

Suppose that $f(a) = g(a) = 0$. Since $f(x)$ and $g(x)$ are differentiable and $g'(a) \neq 0$, then we have

$$\lim_{x \to a} \frac{f(x)}{g(x)} = \lim_{x \to a} \frac{f(x)-f(a)}{g(x)-g(a)} = \frac{\lim_{x \to a} \frac{f(x)-f(a)}{x-a}}{\lim_{x \to a} \frac{g(x)-g(a)}{x-a}} =$$

$$\frac{f'(a)}{g'(a)} = \lim_{x \to a} \frac{f'(x)}{g'(x)}. \quad *$$

The story behind L'Hospital's Rule is intriguing to mathematical scholars and historians alike. A Swiss mathematician named John Bernoulli discovered L'Hospital's Rule in the late 17th century, but the theorem was named after the French nobleman Marquis de L'Hospital. Historians claim that L'Hospital and Bernoulli made a business arrangement. L'Hospital purchased the rights of the theorem from Bernoulli, which is named in honor of him.

CAUCHY'S MEAN VALUE THEOREM

Even though we are not going to prove the general version of L'Hospital's Rule, we can examine an important statement that can help prove the theorem. This statement is known as Cauchy's Mean Value Theorem, which is named in honor of the 19th century French mathematician Augustin-Louis Cauchy.

CAUCHY'S MEAN VALUE THEOREM: If $f(x)$ and $g(x)$ are continuous on $[a, b]$ and differentiable on (a, b) and $g'(x) \neq 0$ for all x in (a, b), then there exists a number c in (a, b) such that

$$\frac{f'(c)}{g'(c)} = \frac{f(b) - f(a)}{g(b) - g(a)}.$$

The proof will be left as an exercise, but is similar to the Mean Value Theorem.

NEWTON'S METHOD

There are methods we have learned in algebra to solve for the zeros (roots) of non-linear functions. For example, we learned how to solve quadratic equations by factoring, completing the square and quadratic formula. For degree 3 and higher equations, we can use factoring in conjunction with synthetic division to find possible zeros. We consult technological devices such as a graphing calculator or computer programs such as MATLAB, Mathematica, etc. to estimate the zeros of non-linear functions.

But is there a method we can utilize by hand to solve non-linear equations? The answer is yes! It is called Newton's Method.

Here is how Newton's Method works. Suppose we let r be a root of a non-linear equation. We can choose a value x_1 to be a guess for r. For the point $(x_1, f(x_1))$ we can draw a tangent line until it crosses the x-axis. The x value that the tangent line touches will be called x_2.

To find the value of x_2 in terms of x_1 we use the point-slope formula from algebra, which is $y - y_1 = m(x - x_1)$. Now we let $m = f'(x)$, $x = x_2$, $y_1 = f(x_1)$, and $y = 0$ since the tangent line touches the point $(x_2, 0)$ on the x-axis.

This becomes $0 - f(x_1) = f'(x)(x_2 - x_1)$ or $-f(x_1) = f'(x)(x_2 - x_1)$, which implies that

$$x_2 - x_1 = \frac{-f(x_1)}{f'(x_1)}.$$ Solving for x_2, we get

$$x_2 = x_1 - \frac{f(x_1)}{f'(x_1)}.$$

Now we can repeat this process to get the next point x_3. We replace x_1 with x_2 and draw the tangent line from the point $(x_2, f(x_2))$ to the x-axis to get the new point x_3. Using the previous formula, we have

$$x_3 = x_2 - \frac{f(x_2)}{f'(x_2)}.$$

If we repeat this process many times, we will come up with the iterative formula

$$x_{n+1} = x_n - \frac{f(x_n)}{f'(x_n)}.$$

This iterative formula can be treated like a sequence.

If $\displaystyle \lim_{n \to \infty} x_n = r,$ then the solution for $f(x) = 0$ converges to r.

The question is how many iterations do we need to make until we get a good approximation? The answer is when a few values of x are almost identical to a few decimal places. Otherwise, we need to generate more iterations until a good solution is found.

There are a few conditions where Newton's Method may fail when approximating zeros. First of all, if $f'(x_n) = 0$, then the value of x_{n+1} is undefined. Secondly, if $f'(x_n) \approx 0$, then the approximation may lie outside the domain of the function, which yields an unreasonable solution. Thirdly, a poor initial guess may be chosen, which results in several iterations that leads to no convergence of a zero.

Overall, Newton's Method is a good way of approximating zeros of non-linear functions, but we notice there are a few drawbacks to the method. The best way to use Newton's Method is to make sure we have a good initial guess, and that our function $f(x)$ is differentiable where $f'(x) \neq 0$ and $f'(x)$ is not close to 0, which will prevent us from getting a value far from the actual zero.

ANTI-DERIVATIVES

Suppose we are given the derivative of a function, and we want to find the original function, then how do we go about doing this? The process of working from the derivative of a function back to the original function is known as anti-differentiation. The original function we find from the derivative is known as the anti-derivative. See the definition of anti-derivative below.

ANTI-DERIVATIVE: If $F'(x) = f(x)$ for all x on an interval I, then $F(x)$ is the anti-derivative of $f(x)$.

The question that arises with anti-derivatives is uniqueness. Can we find another function $G(x)$ such that $G'(x) = f(x)$? The answer is yes! For example if $F(x) = x^2 + 1$ and $G(x) = x^2 - 4$, then by taking the derivatives of $F(x)$ and $G(x)$, we know that

$F'(x) = G'(x) = 2x$. But $F(x) \neq G(x)$ since the two functions differs by a constant. This leads us to the following theorem.

ANTI-DERIVATIVE THEOREM: The general anti-derivative of $f(x)$ on I is $F(x) + C$, where C is a real constant.

PROOF: Let $F'(x) = G'(x) = f(x)$ on I. Then define a function $H(x) = G(x) - F(x)$. Since $F(x)$ and $G(x)$ are differentiable on I, we know that H(x) is also differentiable on I.

So, $H'(x) = G'(x) - F'(x) = f(x) - f(x) = 0$. Since $H'(x) = 0$, this implies that $H(x)$ is a constant function or $H(x) = C$ where C is a real constant. Then we have $H(x) = G(x) - F(x) = C$ or $G(x) = F(x) + C$, which is the anti-derivative of $f(x)$ on I. *

In order to find the general anti-derivative, we need to know the derivative $f(x)$ and find $F(x)$ by process of anti-differentiation and add a constant at the end.

One of the first rules we learned in differentiation is the power rule. Likewise, anti-differentiation also has a power rule.

For example, if $F(x) = \dfrac{x^{n+1}}{n+1}$, where $n \neq -1$ then

$F'(x) = f(x) = x^n$. Thus, the general anti-derivative of x^n is

$\dfrac{x^{n+1}}{n+1} + C$.

Notice that the power rule tells us to add 1 to the power of n and divide by the power $n + 1$. This rule fails for $n \neq -1$ since we get a zero in the denominator. The case for $n = -1$ brings us to the next rule of anti-derivatives.

If $n = -1$, then x^n becomes x^{-1} or $1 / x$. We know that if $F(x) = \ln x$, then $F'(x) = f(x) = 1 / x$. So, the general anti-derivative of x^{-1} or $1 / x$ is $\ln x + C$. This is true for all values of $x > 0$. In addition if we want to consider values of $x < 0$, then our anti-derivative would be $\ln|x| + C$. We include the absolute value sign to account for the negative values of x. Unless the domain is restricted to positive values of x, it is wise to use the latter than the former.

DIFFERENTIAL EQUATIONS

In the past, we have used the rules of algebra to solve algebraic equations in the goal of finding the unknown variable or variables. But we can also use the rules of calculus to solve for an unknown function in an equation that has derivatives. Such equations are called differential equations. At the end of this review, the student will have a chance to solve some basic differential equations. Later on, we will learn more about differential equations and how to solve them by using advanced techniques.

HOW TO SOLVE A BASIC DIFFERENTIAL EQUATION

1. Given the differential equation $F'(x) = f(x)$, use the rules of anti-differentiation to find the general anti-derivative $y = F(x) + C$.

2. Apply the initial condition $F(x_o) = y_o$ by substituting in the value of x_o into x and y_o into y.

3. Solve for the constant C and substitute this value into the solution for y.

4. Check the solution by finding y'. If $y' = f(x)$, then the solution is correct.

   The next review will cover the second problem in calculus, which is finding the area under the curve. This process is known as integration.

COMPUTATIONAL EXERCISES #14

1. Determine if L'Hospital's Rule can be used to evaluate each of the following limits. If a simpler method can be employed, consider using that method. If L'Hospital's Rule cannot be used, explain why.

   a) $\lim_{x \to 0} \dfrac{x^3 + x}{e^x - 1}$

b) $\displaystyle\lim_{x \to \frac{\pi}{2}} \frac{1 - \sin x}{\cos x}$

c) $\displaystyle\lim_{x \to \infty} \frac{\ln x^2}{x}$

d) $\displaystyle\lim_{x \to -2} \frac{x + 2}{x^2 - 4}$

e) $\displaystyle\lim_{x \to \infty} \sqrt{x} - x$

f) $\displaystyle\lim_{x \to 0} \frac{2^x - 1}{5^x - 1}$

2. Evaluate the following limit. Assume that $p > 0$.

$\displaystyle\lim_{x \to \infty} x^{-p} \ln x$

3. Use Newton's Method to estimate the zeros of each function. Round to four decimal places.

a) $f(x) = x^3 + x - 3$

b) $f(x) = x^4 - 15$

4. Estimate each of the following values by using Newton's Method. Round to six decimal places.

a) $\sqrt[3]{7}$

b) $\sqrt[5]{20}$

5. Find an iterative formula to compute $\sqrt{a}$, where $a \geq 0$.

6. Find the general anti-derivative.

a) $4x^2 - 2x + 6$

b) $3\sqrt{x} + \sqrt[7]{x^2}$

c) $\dfrac{8x^{10} + 10x^9 - 5x^8}{x^{10}}$

d) $9e^x + 2\cos x - \dfrac{1}{x}$

7. Find the anti-derivative that satisfies the given initial condition.

   a) $f(x) = 12x^3 - 6x^2$, $F(0) = 3$

   b) $f(x) = 4\sec^2 x + 3\sin x - \cos x$, $F(\pi) = -\pi/2$

8. Find $f(x)$.

   a) $f\,'(x) = x^4 - x^3 + 1$

   b) $f\,'(x) = 7e^x + 2x$, $F'(0) = 1$, $F(0) = -1$

9. Find the position function $s(t)$ given that the acceleration function is $a(t) = -9.8$ m$/$s^2 with initial conditions $v(0) = 25$ m$/$s and $s(2) = 15$ m.

10. The velocity of a particle is the formula $v(t) = 10 - 2t$ ft$/$s, where $s(0) = 20$ ft.

   a) What is the particle's position when it is at rest?

   b) When does the particle change direction?

   c) What is the acceleration of the particle? Is it speeding up or slowing down?

CONCEPTUAL EXERCISES #14

1. True or False: L'Hospital's Rule can be used to find the limit of the quotient of $f(x)$ and $g(x)$ as long as you have an indeterminate form such as $0 / 0$ or ∞ / ∞.

2. True or False: $\lim\limits_{x \to 0} \dfrac{e^x}{x} = 1$

3. True or False: $\lim\limits_{x \to \infty} x^{-2} \ln x = 0$

4. True or False: Newton's Method can be used to solve the equation $x = \cos x$.

5. True or False: The general anti-derivative of $f(x) = \csc^2 x$ is $F(x) = \cot x + C$.

6. True or False: The anti-derivative of $f(x) = x^{-4}$ when $F(1) = 0$ is $F(x) = \dfrac{-1}{3x^3} + \dfrac{1}{3}$.

7. True or False: If $F'(x) = G'(x)$, then $F(x) = G(x)$.

8. True or False: The general anti-derivative of e^x is $e^x + C$.

9. True or False: Newton's Method can possibly fail if $f'(x) \approx 0$.

10. True or False: If $f'(x) > 0$ on an interval I, then $f(x) > 0$ on I.

11. If $f(x) = \dfrac{x^p}{x^q}$, where p and q are positive integers and $q > p$, then show that $\lim\limits_{x \to \infty} f(x) = 0$.

12. Prove Cauchy's Mean Value Theorem. Hint: Define a function

$$h(x) = f(x) - f(a) - \frac{f(b)-f(a)}{g(b)-g(a)}[g(x)-g(a)] \text{ and use}$$

the hypothesis of Rolle's Theorem.

13. For any natural number n, show that $\lim_{x \to \infty} \frac{e^x}{x^n} = \infty$.

When comparing the exponential function e^x and the power function x^n, what does mean when the value of x gets really large?

14. Consider the acceleration function of a projectile $a(t) = -g$, where g is the acceleration due to gravity. If $v(0) = v_o$(initial velocity) and $s(0) = s_o$, (initial position) then show that the position function is $s(t) = s_o + v_o t - \frac{1}{2}gt^2$.

15. The equation $s(t)$ from exercise 14 can be broken down into x and y components. The height of the projectile can be described by $y(t) = y_o + v_o \sin\theta t - \frac{1}{2}gt^2$ and the range of the projectile can be described by $x(t) = v_o \cos\theta t$, where y_o is the initial height of the projectile and θ is the launch angle respect to the ground.

a) If the projectile is launched from the ground, find the time when the projectile reaches the maximum height and find the maximum height.

b) At what time does the projectile hit the ground? What is the range of the projectile?

c) At what angle should one launch the projectile to achieve maximum range?

REVIEW #15

CALCULUS 1 PART 5

In this review, we are going to investigate the second problem that calculus attempts to answer, which is finding the area under the curve. From algebra, we know how to calculate the area of simple geometric figures such as a square, rectangle, triangle, etc. But how do we calculate the area of under the curve, which is not so simple?

THE DEFINITE INTEGRAL

Suppose we have a continuous function $f(x)$ on $[a, b]$ and let R be the shaded region underneath the graph of $f(x)$. One way we can estimate the area under the curve is to break R into n number of rectangles with equal widths.

If we let Δx represent the width of each rectangle and $f(x_i^*)$ be the height of the rectangle chosen at a sample point x_i, then we can find the area to be

$$f(x_1^*)\,\Delta x + f(x_2^*)\,\Delta x + \ldots + f(x_n^*)\,\Delta x, \text{ where } \Delta x = \frac{b-a}{n}.$$

Now each sample point x_i^* is selected from each of the sub-intervals $[x_0, x_1], [x_1, x_2], \ldots, [x_{n-1}, x_n]$, where

$$x_0 = a, x_1 = a + \Delta x, x_2 = a + 2\Delta x, \ldots, x_n = a + n\Delta x.$$

If we decrease the width of each rectangle, then we can fit even more rectangles under the curve and receive a better estimation of the area. Now suppose that we can fit an infinite number of rectangles under the curve. Then the area under the curve can be expressed as

$$A = \lim_{n \to \infty} \sum_{i=1}^{n} f(x_i^*)\Delta x.$$

Now we can get estimations of the area under the curve by letting the top right corner of each rectangle touching the curve or letting the top left corner of each rectangle touching the curve. These estimations are known as the right endpoints and left endpoints respectively.

RIGHT END POINTS: $R_n = \displaystyle\lim_{n \to \infty} \sum_{i=1}^{n} f(x_i)\Delta x$

LEFT END POINTS: $L_n = \displaystyle\lim_{n \to \infty} \sum_{i=1}^{n} f(x_{i-1})\Delta x$

There is another way we notate the area under the curve without using the summation symbol. It is called the definite integral.

DEFINITE INTEGRAL: If $f(x)$ is continuous on $[a, b]$, then we can divide $[a, b]$ into sub-intervals of equal width $\Delta x = \dfrac{b-a}{n}$, where $[x_0, x_1]$, $[x_1, x_2]$, ..., $[x_{n-1}, x_n]$ are the sub-intervals and x_1^*, x_2^*, ..., x_n^* are the sample points taken in the respective intervals, then

$$A = \lim_{n \to \infty} \sum_{i=1}^{n} f(x_i^*)\Delta x = \int_{a}^{b} f(x)dx .$$

$\displaystyle\int_{a}^{b} f(x)dx$ can be read as "the integral from a to b of $f(x)$ dx." The numbers a and b are called the limits of integration.

The number a is the lower limit of integration. The number b is the upper limit of integration. The function $f(x)$ is the integrand or the function we are integrating. The dx term represents a differential or the variable we are integrating $f(x)$ respect to. The process of finding the integral of a function $f(x)$ is called integration.

It is important to note that when we evaluate a definite integral, we get a number as a result. The number does not depend on the independent variable of our function.

Another rule we can use is to take the middle number $x_i{'}$ or average number between the endpoints of all the sub-intervals $[x_{i-1}, x_i]$. This is known as the Midpoint Rule.

MIDPOINT RULE: $\displaystyle\int_a^b f(x)dx \approx \sum_{i=1}^{n} f(x_i')\Delta x$,

where $x_i' = \dfrac{x_{i-1} + x_i}{2}$.

The sum $\displaystyle\sum_{i=1}^{n} f(x_i{}^*)\Delta x$ has an important name, which is called the Riemann sum named after the 19th century German mathematician Bernhard Riemann. The Riemann sum can be thought of as the sum of the areas of rectangles provided that $f(x) \geq 0$.

If the definite integral gives us a positive number, then we can interpret this number as the area under the curve for $f(x)$. On the other hand, if the definite integral gives us a negative number, then we cannot interpret it as the area under the curve since area is considered to be a non-negative number.

Bernhard Riemann was a 19th century German mathematician who received his Ph.D. under the advisement of Carl F. Gauss. Though Gauss did not usually give praise or accolades to many of his students, he did give favor to Riemann for his achievements. Gauss believed that Riemann was one of the most creative, talented mathematicians in his day. Riemann made important contributions to calculus and geometry, which laid the foundation of Albert Einstein's general theory of relativity. Unfortunately, Riemann lived an ephemeral life dying of tuberculosis at the age of 39.

SUMMATION PROPERTIES

There are a few summation properties that we need to know in order to evaluate the definite integral by definition. In conjunction to these properties, we should use the properties from the conceptual exercises in Review #4.

1. $\displaystyle\sum_{i=1}^{n} i = \frac{n(n+1)}{2}$

2. $\displaystyle\sum_{i=1}^{n} i^2 = \frac{n(n+1)(2n+1)}{6}$

3. $\displaystyle\sum_{i=1}^{n} i^3 = \left(\frac{n(n+1)}{2}\right)^2$

PROPERTIES OF THE DEFINITE INTEGRAL

Here are some useful properties when evaluating definite integrals. Assume that both functions $f(x)$ and $g(x)$ are continuous on $[a, b]$.

OPPOSITE PROPERTY: $\displaystyle -\int_a^b f(x)dx = \int_b^a f(x)dx$

ZERO PROPERTY: If $a = b$, then $\displaystyle\int_a^b f(x)dx = 0$.

CONSTANT PROPERTY: If c is any real number, then $\displaystyle\int_a^b cdx = c(b-a)$.

SUM/DIFF PROPERTY: $\displaystyle\int_a^b [f(x) \pm g(x)]dx = \int_a^b f(x)dx \pm \int_a^b g(x)dx$

DECOMPOSITION PROPERTY: If c is a number between a and b

and $f(x)$ is continuous on $[a, c]$ and $[c, b]$, then

$\displaystyle\int_a^b f(x)dx = \int_a^c f(x)dx + \int_c^b f(x)dx$.

The opposite property says that if we want to find the opposite of the definite integral of $f(x)$ from a to b, we find the definite integral of $f(x)$ from b to a. We just switch the limits of integration.

The zero property is self-explanatory. Since $a = b$, this implies that the width of the rectangle is zero. Thus, there is no area.

The constant property says that if we have a rectangle of height c, and a width of $b - a$, then the area of the rectangle is width x height or $c(b - a)$.

The sum and difference property says that the definite integral of the sum or difference of two functions is the sum or difference of the two definite integrals.

The decomposition property tells us that we can divide the interval $[a, b]$ into two intervals $[a, c]$ and $[c, b]$ and find the area under the curve of $f(x)$ for each of these two intervals and add their results together.

COMPARISON PROPERTIES

When we compare one function to another function, we can also compare how their areas will look. Below are three important comparison properties of definite integrals.

POSITIVE/ZERO PROPERTY: If $f(x) \geq 0$ for $[a, b]$, then $\int_a^b f(x)dx \geq 0$.

FUNCTION COMPARSION PROPERTY: If $f(x) \geq g(x)$ for $[a, b]$, then $\int_a^b f(x)dx \geq \int_a^b g(x)dx$.

MINIMUM/MAXIMUM PROPERTY: If m and M are real numbers where $m \leq f(x) \leq M$ for $[a, b]$, then $m(b-a) \leq \int_a^b f(x)dx \leq M(b-a)$.

The positive/zero property says that if $f(x)$ is either positive or zero for all x in $[a, b]$, then either the area under $f(x)$ will either be positive or zero.

The function comparison property compares two functions $f(x)$ and $g(x)$. If $f(x)$ is greater than $g(x)$ for all x on $[a, b]$, then $f(x)$ will have a larger area than $g(x)$. If $f(x) = g(x)$ for all x on $[a, b]$, then both $f(x)$ and $g(x)$ will have equal areas.

The minimum/maximum property helps us to find an estimate for the area under the curve, especially for non-elementary functions whose integrals are difficult to evaluate. If $f(x)$ lies between a minimum m and a maximum M, then the area under $f(x)$ will lie between $m(b-a)$ and $M(b-a)$. We can think of the area of $f(x)$ is "sandwiched" between the area of $m(b-a)$ (smaller area with height m and width $b-a$) and the area of $M(b-a)$ (larger area with height M and width $b-a$).

Now we are going to look at an important theorem that ties both the tangent line problem and the area problem together.

FUNDAMENTAL THEOREM OF CALCULUS

The Fundamental Theorem of Calculus is divided into two parts. Let us look at the first part.

PART 1: Let $f(x)$ be continuous on $[a, b]$ and define the function

$g(x) = \int_a^x f(t)dt$. Then $g(x)$ is continuous on $[a, b]$ and differentiable on (a, b) and $g'(x) = f(x)$.

We will not give a formal proof here, but we can provide some insight to see why this is true.

Suppose that $f(x)$ is continuous on $[a, b]$ and assume that $f(x) \geq 0$ and that $h > 0$. From the graph below, we can see that $g(x + h) - g(x) \approx hf(x)$. Or we can express this as $f(x) \approx \dfrac{g(x+h) - g(x)}{h}$. By the taking the limit as h approaches 0, it follows that $\lim_{h \to 0} \dfrac{g(x+h) - g(x)}{h} = f(x)$. But the left side of the equation indicates the derivative of $g(x)$ or $g'(x)$. So, we can conclude that $g'(x) = f(x)$.

F, T, C, Part 1

$$\text{If } g(x) = \int_a^x f(t)\, dt, \text{ then } g'(x) = f(x).$$

The second part of the Fundamental Theorem of Calculus is a direct consequence of the first part. Let us have a look at the second part.

PART 2: If $f(x)$ is continuous on $[a, b]$, then $\int_a^b f(x)dx = F(b) - F(a)$, where $F'(x) = f(x)$. That is, $F(x)$ is the anti-derivative of $f(x)$.

To see why this is true, suppose that $F(x) = g(x) + C$, where the function $g(x) = \int_a^x f(t)dt$ from the first part of the Fundamental Theorem of Calculus. Now $F(b) = g(b) + C$ and $F(a) = g(a) + C$. Then we have

$$F(b) - F(a) = g(b) + C - (g(a) + C) = g(b) - g(a) = \int_a^b f(t)dt - \int_a^a f(t)dt =$$

$$\int_a^b f(t)dt - 0 = \int_a^b f(t)dt.$$

So, the second part of the Fundamental Theorem of Calculus says that we find the area under the curve of $f(x)$ by evaluating $F(x)$ at the endpoints of the closed interval, and find the difference of these two values.

Now what if we want to find the general anti-derivative of a function without involving a closed interval? This type of integral is called an indefinite integral.

INDEFINITE INTEGRAL: Let $F'(x) = f(x)$, where $F(x)$ is the anti-derivative of $f(x)$. Then the indefinite integral can be defined as

$\int f(x)dx = F(x) + C$. The real number C is known as the constant of integration.

Notice that with an indefinite integral, we do not have limits of integration. We are only interested in finding the general function. The value of C added to the function indicates that there a family of functions that are possible. If we want to find the value of the constant, we need to know the initial condition $F(x_o) = y_o$ to plug into the function and solve for C.

COMMONLY USED INDEFINITE INTEGRALS

Below are fifteen commonly used indefinite integrals that the student should get acquainted with in their first year of calculus.

1. For any constant c, $\int cf(x)dx = c\int f(x)dx$.

2. For any constant k, $\int k\,dx = kx + C$.

3. $\int x^n dx = \dfrac{x^{n+1}}{n+1} + C$, where $n \neq -1$.

4. $\int e^x dx = e^x + C$

5. $\int \dfrac{1}{x}dx = \ln|x| + C$

6. $\int a^x dx = \dfrac{a^x}{\ln a} + C$

7. $\int \cos x\,dx = \sin x + C$

8. $\int \sin x\,dx = -\cos x + C$

9. $\int \sec^2 x \, dx = \tan x + C$

10. $\int \csc^2 x \, dx = -\cot x + C$

11. $\int \sec x \tan x \, dx = \sec x + C$

12. $\int \csc x \cot x \, dx = -\csc x + C$

13. $\int \dfrac{1}{\sqrt{1-x^2}} \, dx = \sin^{-1} x + C$

14. $\int \dfrac{-1}{\sqrt{1-x^2}} \, dx = \cos^{-1} x + C$

15. $\int \dfrac{1}{1+x^2} \, dx = \tan^{-1} x + C$

Although these integrals are good to use in general, we need to learn some important techniques or rules to evaluate integrals in other forms that may not look so familiar. One of the first rules we will learn and use is called u-substitution.

U-SUBSTITUTION RULE

Suppose that $u = g(x)$ is a differentiable formula, where the range of $g(x)$ is an interval I. And let $f(x)$ be continuous on I. Then the u-substitution rule is defined as $\int f(g(x))g'(x)dx = \int f(u)du$. The proof of the u-substitution rule will be left as an exercise.

With the u-substitution rule, we work with both dx and du and treat them as differentials. Finding the correct substitution for u can be a little difficult at first, but with patience, perseverance, and a little practice, one can master this art form. Now if $u = g(x)$ does not work, then we pick another function for u and repeat this process until the substitution for u does work.

The best way to determine u is to see if the du expression (found from the derivative of u) is in the numerator. If du is in the numerator, then our choice for u is a good one.

We can also apply the *u*-substitution for definite integrals. Let us observe how this works.

U-SUBSTITUTION FOR THE DEFINITE INTEGRAL: If $g'(x)$ is continuous on $[a, b]$ and $f(x)$ is continuous on the range of $u = g(x)$, then

$$\int_a^b f(g(x))g'(x)dx = \int_{g(a)}^{g(b)} f(u)du \, .$$

PROOF: Since $g'(x)$ is continuous on $[a, b]$ and $f(x)$ is continuous on the range of $u = g(x)$, we can evaluate the definite integral.

Evaluating the left hand side of the equation, we have

$$\int_a^b f(g(x))g'(x)dx = F(g(x))\big|_a^b = F(g(b)) - F(g(a)).$$

Evaluating the right hand side of the equation, we have

$$\int_{g(a)}^{g(b)} f(u)du = F(u)\big|_{g(a)}^{g(b)} = F(g(b)) - F(g(a)).$$

Thus, both the left hand and right hand sides of the equation are equal. *

There are a couple more properties we need to look at before wrapping up this review. They are the symmetry properties.

SYMMETRY PROPERTIES

Let $f(x)$ be continuous on $[-a, a]$. Then we have the following.

a) If $f(x)$ is even, then $\int_{-a}^{a} f(x)dx = 2\int_{0}^{a} f(x)dx$.

b) If $f(x)$ is odd, then $\int_{-a}^{a} f(x)dx = 0$.

PROOF:

a) We can rewrite the left hand side of the equation as

$$\int_{-a}^{a} f(x)dx = \int_{-a}^{0} f(x)dx + \int_{0}^{a} f(x)dx \text{ by the decomposition property of}$$

integrals. Now if we let $u = -x$, which means $x = -u$ and $dx = -du$, then
we have

$$\int_{-a}^{0} f(-u)(-du) = -\int_{0}^{a} f(-u)(-du) = \int_{0}^{a} f(-u)du \text{ by the opposite property}$$

of integrals and that $u = a$. Since $f(x)$ is even, then we have
$f(-u) = f(u)$.

So, $\int_{-a}^{a} f(x)dx = \int_{0}^{a} f(u)du + \int_{0}^{a} f(x)dx = 2\int_{0}^{a} f(x)dx.$ *

Notice that u, x, du and dx can be used interchangeably since the value
obtained from the definite integral is independent of the variable
chosen.

b) We can use the formula from part a) to get

$$\int_{-a}^{a} f(x)dx = \int_{0}^{a} f(-u)du + \int_{0}^{a} f(x)dx.$$

But since $f(x)$ is odd, we know that $f(-u) = -f(u)$. Thus we have

$$\int_{-a}^{a} f(x)dx = -\int_{0}^{a} f(u)du + \int_{0}^{a} f(x)dx = 0.$$

COMPUTATIONAL EXERCISES #15

1. Use the definition of the integral to compute $\int_{0}^{2} x^2 dx$.

2. Use the left-end points to estimate $\int_{1}^{3} \ln x dx$ for $n = 5$.

3. Use the right-end points to estimate $\int_{2}^{6} (\sqrt{x} + x)dx$ for $n = 4$.

214

4. Evaluate the limit. $\displaystyle\lim_{n \to \infty} \sum_{i=1}^{n} \frac{4i}{n^2}$.

5. Evaluate each of the following indefinite integrals.

a) $\displaystyle\int 5dx$

b) $\displaystyle\int x(3 - x^4)dx$

c) $\displaystyle\int (e^x + 2\cos x - \sin x)dx$

d) $\displaystyle\int \frac{8}{1 + x^2} dx$

6. Evaluate each of the following definite integrals.

a) $\displaystyle\int_{5}^{10} \frac{4 - x}{x} dx$

b) $\displaystyle\int_{-1}^{1} (x^3 + 2x)dx$

c) $\displaystyle\int_{0}^{\frac{\pi}{4}} \sec x(\sec x + \tan x)dx$

7. Use u-substitution to evaluate each of the following integrals.

a) $\displaystyle\int 9x^2 \sqrt{x^3 + 8}dx$

b) $\displaystyle\int_{0}^{7} 6e^{3x} dx$

c) $\displaystyle\int \frac{\sin(\ln x)}{x} dx$

d) $\displaystyle\int_{2}^{5} \frac{x + 1}{x - 1} dx$

8. Use the Fundamental Theorem of Calculus to find $g'(x)$.

    a) $g(x) = \int_0^x e^{-t^2} dt$

    b) $g(x) = \int_0^{x^3} \tan t\, dt$

9. Estimate a lower and upper bound for the area. $\int_{-1}^{1} xe^x dx$

10. If $\int_1^3 f(x)dx = \ln 6$ and $\int_2^3 f(x)dx = \ln 3$, then what is $\int_1^2 f(x)dx$?

CONCEPTUAL EXERCISES #15

1. True or False: $\int_a^b f(x)dx = \lim_{n \to \infty} \sum_{i=1}^{n} f(x_i)\Delta x$, where $\Delta x = \dfrac{b-a}{n}$.

2. True or False: In order to get a better accuracy of the area under the curve, you need to fit more rectangles under the curve.

3. True or False: $\int_{-3}^{3} x^4 dx = 0$.

4. True or False: $\int_a^b c\,dx$ can be interpreted as the area of a rectangle with a width of $b - a$ and a height of c.

5. True or False: If $u = g(x)$, then $\int_a^b f(g(x))g'(x)dx = \int_{f(a)}^{f(b)} g(u)du$.

6. True or False: If $a = b$, then $\int_a^b f(x)dx = 0$.

7. True or False: If $f(x)$ is differentiable on (a, b) and $f(x)$ is continuous on $[a, b]$, then $\int_a^b f(x)dx$ exists.

8. True or False: If $f(x)$ is an odd function, k is a non-zero constant, and $\int_a^b f(x)dx = \dfrac{1}{k}$ then $\int_b^a kf(-x)dx = 1$

9. True or False: If $f(x)$ is continuous and increasing on $[a, b]$, then $\int_a^b f(x)dx > 0$.

10. True or False: $\int \csc^2 x = \cot x + C$

11. Use the definition of the integral to show that $\int_0^1 x^2 dx = \dfrac{1}{3}$.

12. Use mathematical induction to prove that $\sum_{i=1}^n i^2 = \dfrac{n(n+1)(2n+1)}{6}$.

13. Prove the opposite, zero, and constant properties of the definite integral.

14. If $x > 0$, then show that $\int_0^x e^{-t}dt \geq \int_0^x e^{-t^2} dt$.

15. Prove that $\left| \int_0^{\sqrt{\pi}} \cos x^2 dx \right| \leq \sqrt{\pi}$.

16. Prove the U-Substitution Rule for indefinite integrals.

17. Show that $\int_{-c}^c (ax^4 + bx^2)dx = 2\int_0^c (ax^4 + bx^2)dx$.

18. Prove or disprove. If $a > 0$ and x is real number, then $\int \tan(ax)dx = -\dfrac{1}{a}\ln(\cos(ax)) + C.$

19. Prove or Disprove. $\int \sec x dx = \ln|\sec x + \tan x| + C$

REVIEW #16

CALCULUS 2 PART 1

We have finished the review of calculus 1, which is a summary of differential calculus and its applications. Now we are moving forward to review calculus 2, which is a summary of integral calculus. In this first part, we are going to examine some important applications of the integral such as finding the area between curves, finding the volume of a solid, finding the average value of a function, and the Mean Value Theorem of Integrals.

AREA BETWEEN CURVES

Suppose we have continuous functions $f(x)$ and $g(x)$ on $[a, b]$ and both $f(x)$ and $g(x)$ are differentiable on (a, b). Since both $f(x)$ and $g(x)$ are continuous on $[a, b]$ and differentiable, we know that $\int_a^b f(x)dx$ and

$\int_a^b g(x)dx$ exist. What if we want to find the area between the graphs of $f(x)$ and $g(x)$? If we let $f(x) \geq g(x)$ for all x in $[a, b]$, then we know that $\int_a^b f(x)dx \geq \int_a^b g(x)dx$, which implies that $\int_a^b f(x)dx - \int_a^b g(x)dx \geq 0$. This tells us that we can find the difference between the areas under $f(x)$ and $g(x)$. See the figure below.

AREA BETWEEN TWO FUNCTIONS: The area between two continuous, differentiable functions $f(x)$ and $g(x)$ is

$$A = \int_a^b [f(x) - g(x)]dx$$

What are the general steps to finding the area between $f(x)$ and $g(x)$? There are three easy steps to determine the area between their graphs.

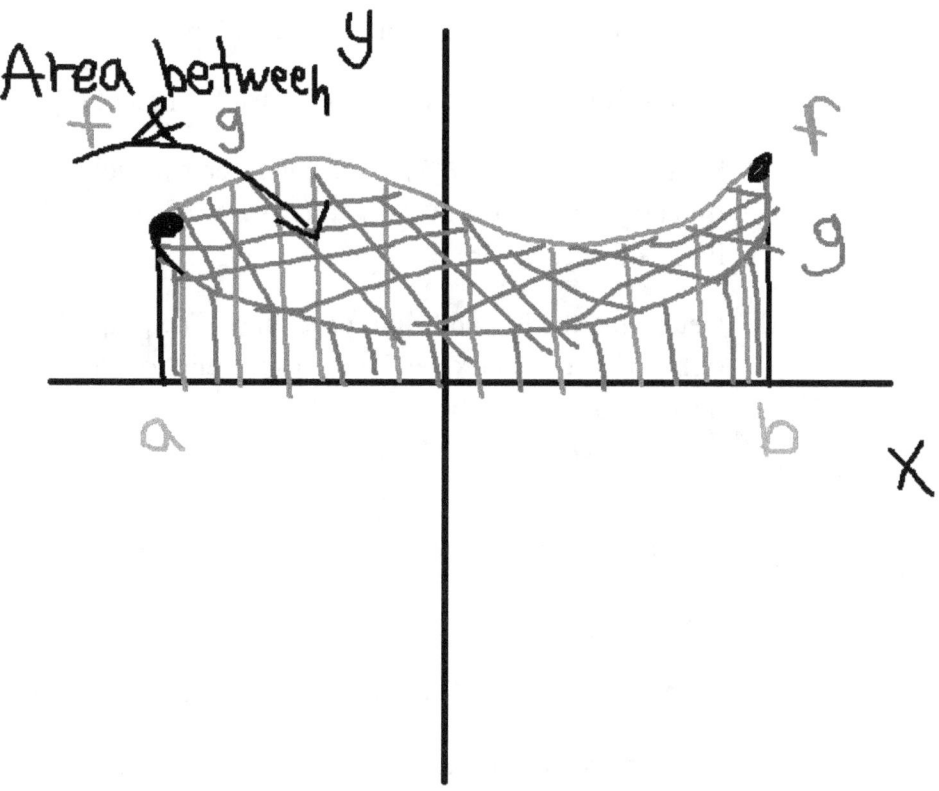

Area between *f* & *g*

STEPS TO FINDING THE AREA BETWEEN *F(X)* AND *G(X)*

1. Find the points where $f(x)$ and $g(x)$ intersect each other. That is, find all x such that $f(x) = g(x)$. These values of x will serve as the limits of integration.

2. Determine the top and bottom functions. If $f(x) > g(x)$ for all x in $[a, b]$, then $f(x)$ is the top function and $g(x)$ is the bottom function. Thus, AREA = TOP FUNCTION – BOTTOM FUNCTION.

3. Evaluate the integral $A = \int_{a}^{b} [f(x) - g(x)]dx$ to find the area between the two curves.

But what if we have the case where we have functions of y instead of x? That is, how do we find the area between the curves of two functions $f(y)$ and $g(y)$? The process is similar, but instead of finding

219

the difference between the top and bottom functions, we find the difference between the right and left functions.

STEPS TO FINDING THE AREA BETWEEN F(Y) AND G(Y)

1. Find the points where $f(y)$ and $g(y)$ intersect each other. That is, find all y such that $f(y) = g(y)$. These values of y will serve as the limits of integration.

2. Determine the left and right functions. The graph farthest to the right $f(y)$ is considered to be the right function, and the graph farthest to the left $g(y)$ is considered to the left function. Thus, the AREA = RIGHT FUNCTION – LEFT FUNCTION.

3. Evaluate the integral $A = \int_{c}^{d} [f(y) - g(y)]dy$ to find the area between the two curves.

See the figure that illustrates the area between $f(y)$ and $g(y)$.

It is important to note that when we use the area = top – bottom, we use the differential dx, and when the area = right – left we use the differential dy.

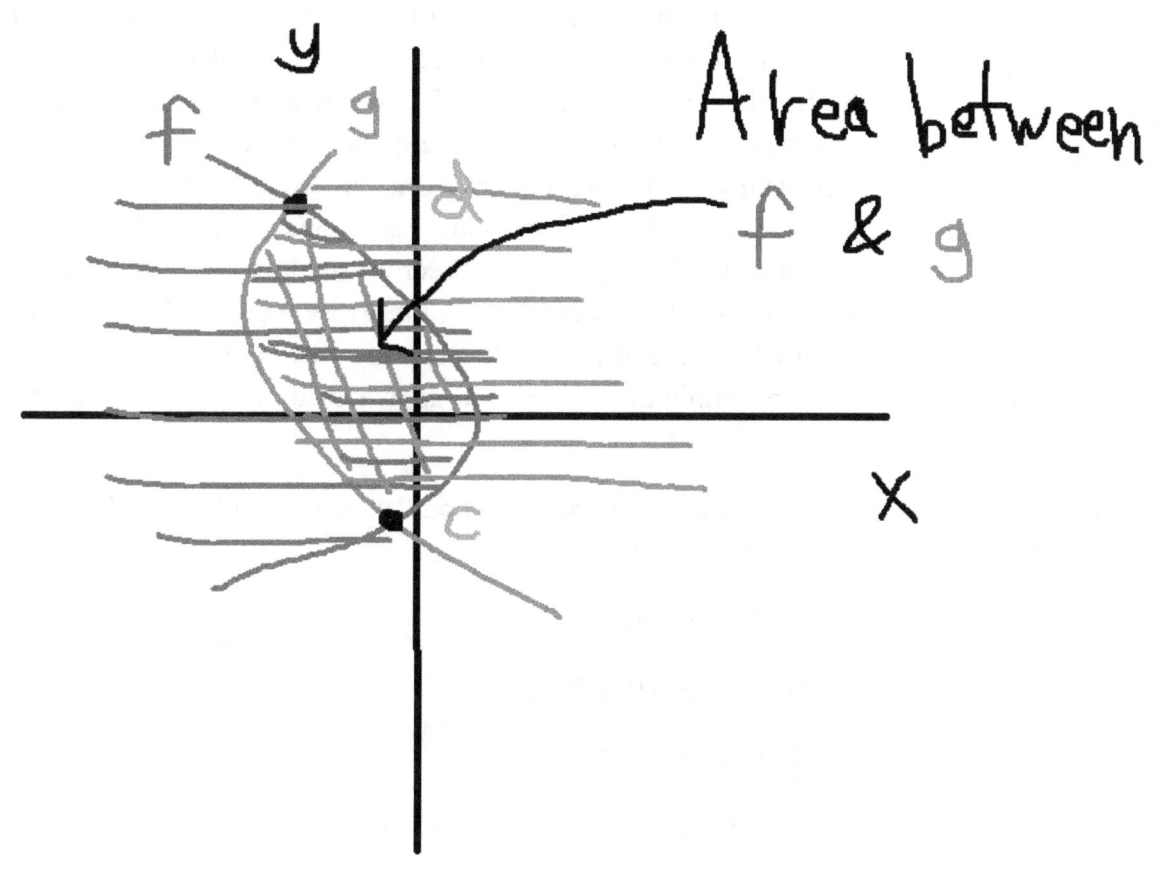

y

f g

d

Area between
f & g

x

c

221

VOLUME OF A SOLID

So far, we have learned how to find the area between the curves of two functions $f(x)$ and $g(x)$ using calculus. We can take this idea a step further to find the volume of a solid. What we mean by a solid, is we can take the graph of a function $f(x)$ and rotate it about the x or y axis to create a three dimensional solid. Now from geometry, we know that the volume $V =$ Area of Base x Height (Thickness). We can slice the solid somewhere near the middle and examine the cross-section. We can add up all of the areas of the cross-sections and multiply by their height or thickness to get the total volume of the solid. But since there are many of these cross-sections within the solid to add up, we need to use the integral to calculate its volume.

There are three important methods we are going to use when finding the volume of a solid.

1. DISK METHOD

2. WASHER METHOD

3. SHELL METHOD

Let us see how each of these methods work.

DISK METHOD

Suppose we have a function $f(x)$ that is continuous on $[a, b]$ and differentiable on (a, b) and let the region R be bounded by $f(x)$, $x = a$, $x = b$, and the x-axis.

If $f(x)$ is rotated about the x-axis, then we can take a vertical slice of $f(x)$, which is perpendicular to the x-axis. The cross-section is a solid disk. The radius $r = f(x)$, while the area of the cross-section is $\pi[f(x)]^2$ with thickness dx.

Therefore, the volume V of the solid generated by rotating $f(x)$ about the x-axis from $x = a$ to $x = b$ is

$$V = \int_a^b \pi[f(x)]^2 \, dx.$$

On the other hand, if we have a function $f(y)$ is continuous on $[c, d]$ and differentiable on (c, d) and let the R be bounded by $f(y)$, $y = c$, $y = d$, and the y-axis.

If $f(y)$ is rotated about the y-axis, then we can take a horizontal slice of $f(y)$, which is perpendicular to the y-axis. The radius $r = f(y)$, while the area of the cross-section is $\pi[f(y)]^2$ with thickness dy.

Therefore, the volume V of the solid generated by rotating $f(y)$ about the y-axis from $y = c$ to $y = d$ is

$$V = \int_c^d \pi[f(y)]^2 \, dy.$$

See the figures below that illustrate the disk method.

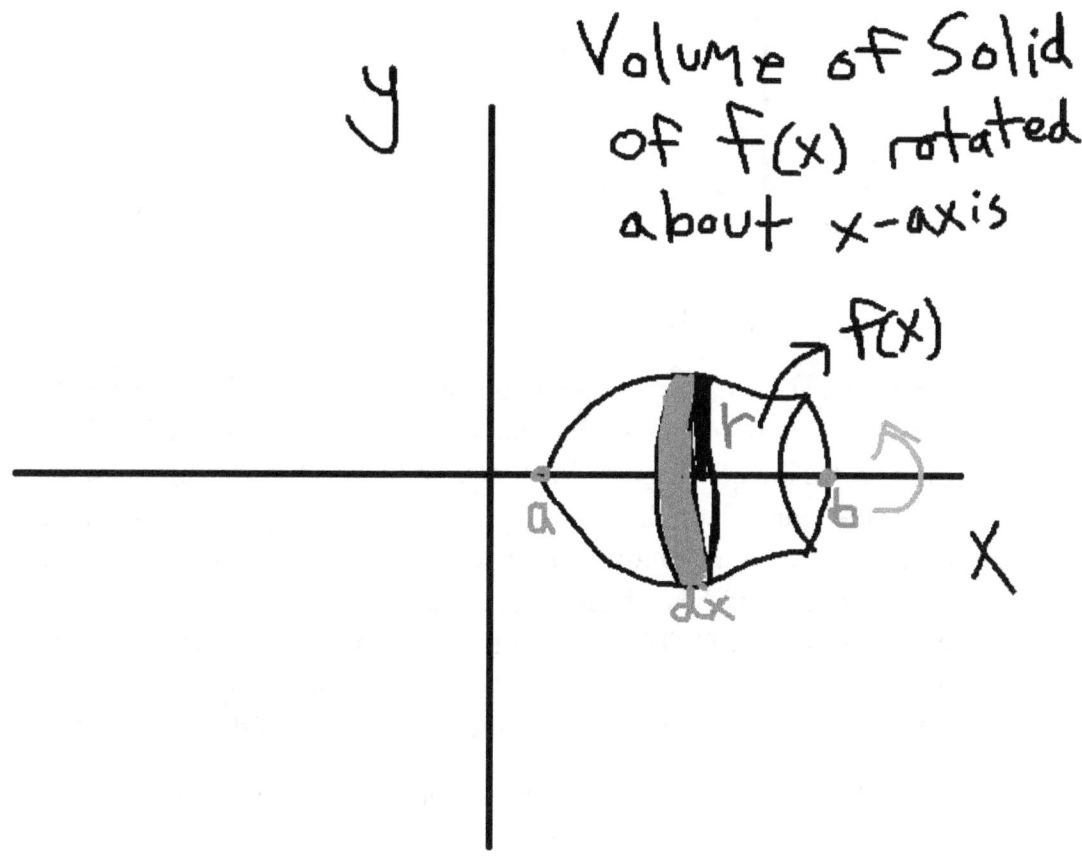

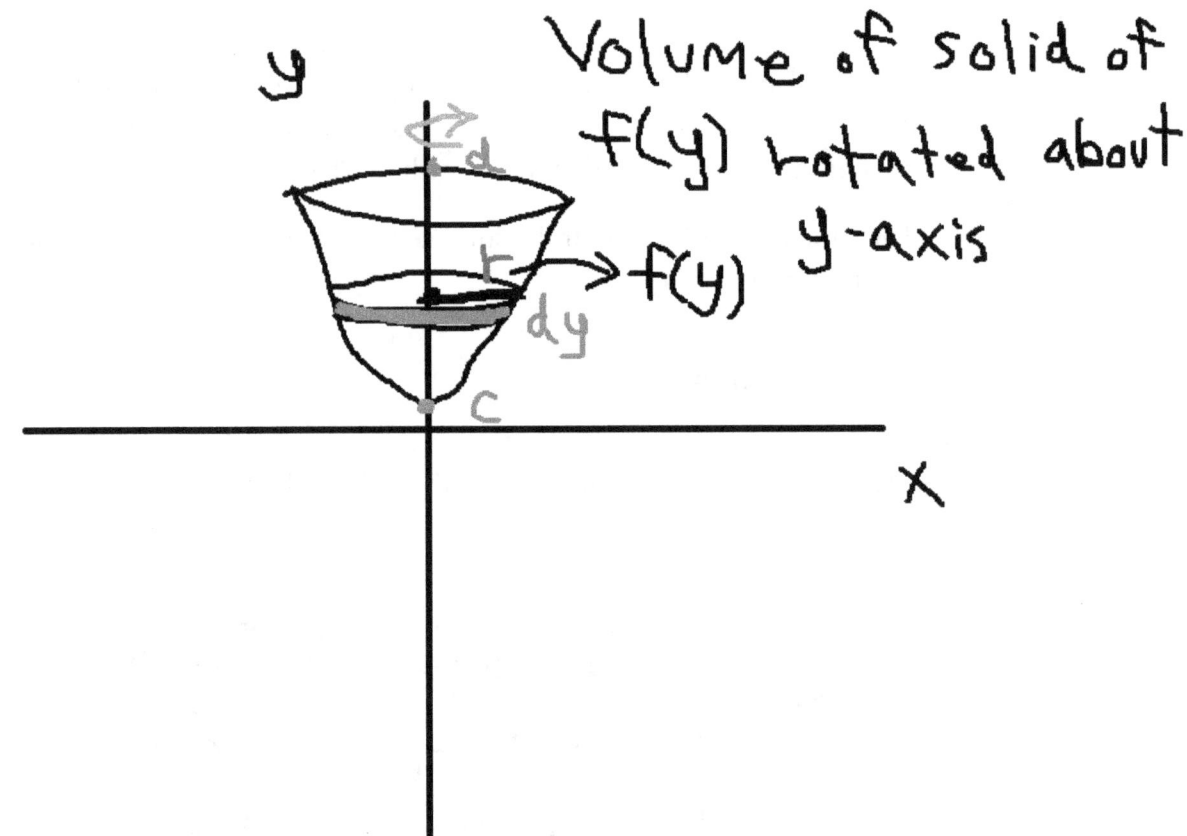

Volume of solid of f(y) rotated about y-axis

WASHER METHOD

Suppose that we have two continuous functions $f(x)$ and $g(x)$ that are continuous on $[a, b]$ and differentiable on (a, b) and $f(x) \geq g(x)$. Then we can find the area of the region R between the two functions. When we rotate the graph about the x-axis, we get a cross-section that is not a solid disk, but something hollow like a washer. Thus, the formula for the cross-sectional area will be different than the disk. We use the formula AREA = TOP – BOTTOM. Each of the washers has a height or thickness dx. Once we add up all of the areas of the individual washers and multiply it by their thickness, then we have the total volume of the generated solid.

If $R = f(x)$ is the outer radius of the washer and $r = g(x)$ is the inner radius of the washer, then the cross-sectional area of the washer is $\pi(R^2 - r^2) = \pi[f(x)^2 - g(x)^2]$ with a thickness dx.

Then the volume V of the solid generated, which is rotated about the x-axis from $x = a$ to $x = b$ is

$$V = \int_a^b \pi[f(x)^2 - g(x)^2]\,dx.$$

On the other hand, we can have two functions $f(y)$ and $g(y)$ that are continuous on $[c, d]$ and differentiable on (c, d) so that $f(y) \geq g(y)$. The outer radius of the washer is $R = f(y)$ and the inner radius of the washer is $r = g(y)$. So, the cross-sectional area of the washer is $\pi[R^2 - r^2] = \pi[f(y)^2 - g(y)^2]$ with a thickness dy. If the graph is rotated about the y-axis, then we can express the volume as

$$V = \int_c^d \pi[f(y)^2 - g(y)^2]\,dy.$$

Notice that the washer method is similar to the disk method in that we slice the graph perpendicular to the axis of rotation. The only difference is their formulas for the cross-sectional areas, which yield different formulas for the volume.

See the figures that illustrate the washer method.

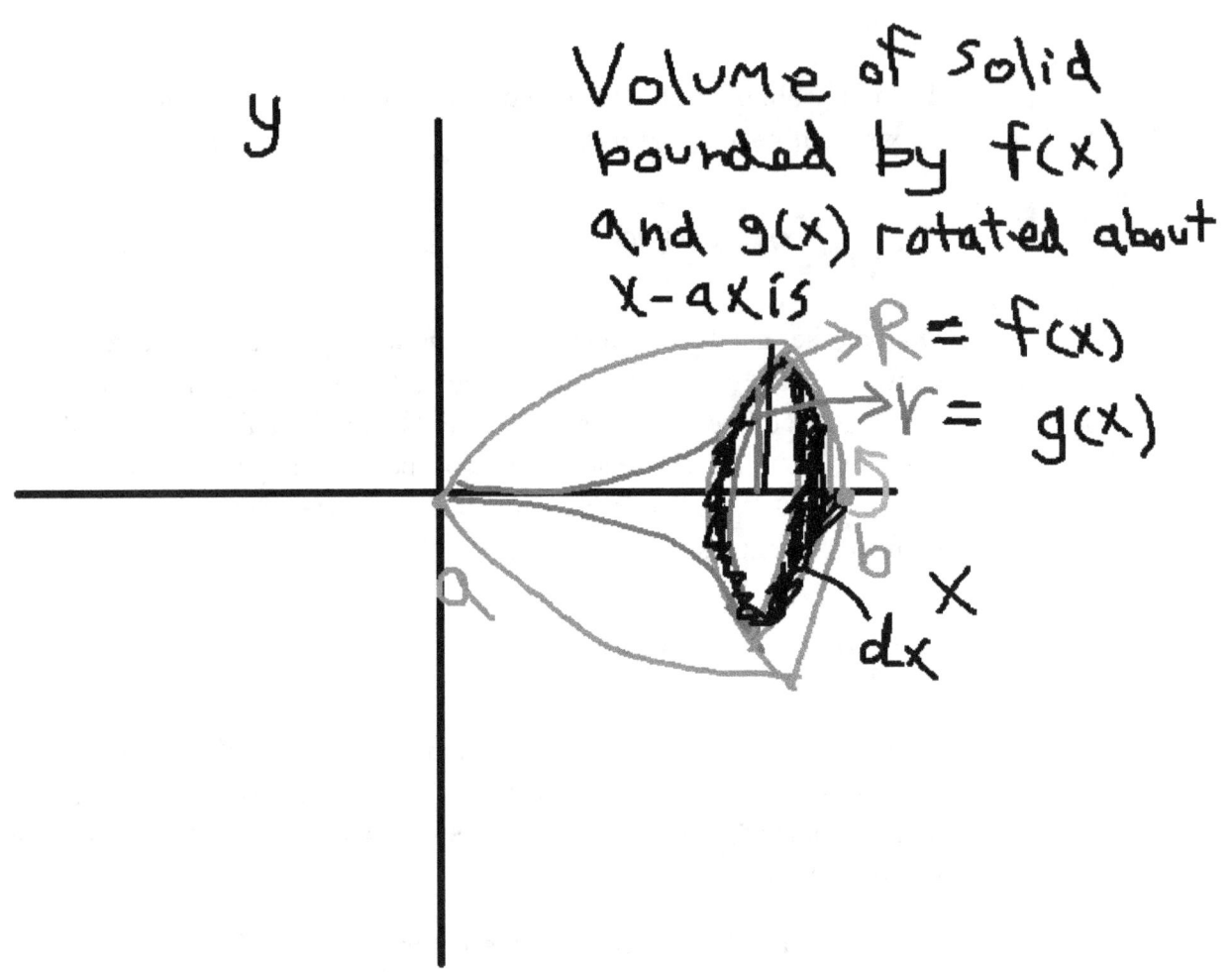

Volume of solid bounded by f(x) and g(x) rotated about x-axis

R = f(x)

r = g(x)

dx

y

x

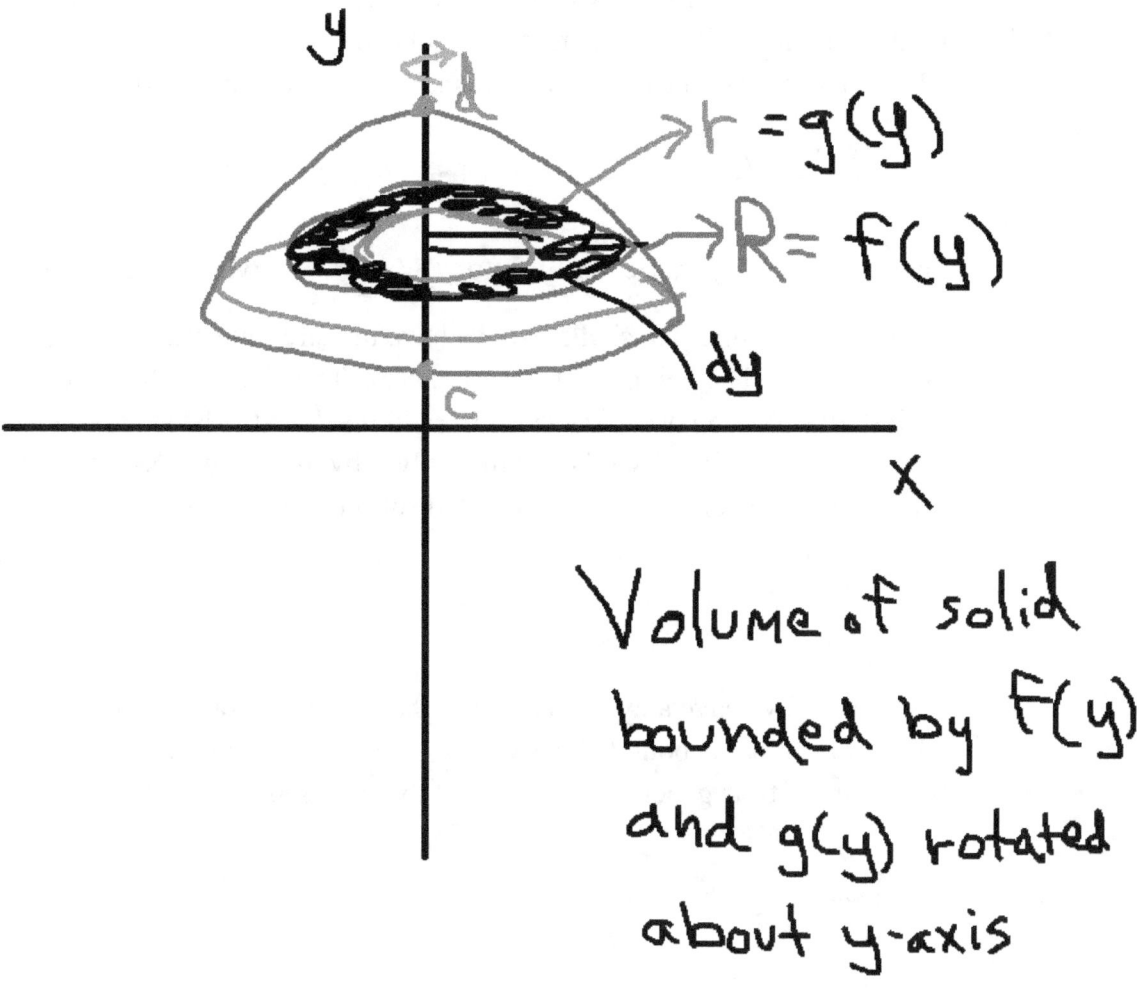

$r = g(y)$

$R = f(y)$

dy

Volume of solid bounded by $f(y)$ and $g(y)$ rotated about y-axis

SHELL METHOD

The last method we will discuss on finding the volume of a generated solid is the shell method. The last two methods show us how to find volumes of solids through solid disks and washers. The shell method allows us to core or peel layers through the solid like a piece of fruit. The cross-section we obtain when we use the shell method is a cylindrical shell.

Suppose that we have a continuous function $f(x)$ on $[a, b]$ and differentiable on (a, b). We let the region R be bounded by $f(x)$, $x = a$, $x = b$, and the x-axis.

From the washer method, we know that the cross-sectional area of the washer is $\pi(R^2 - r^2)$. (Recall that $R = f(x)$ and $r = g(x)$). Now $R^2 - r^2 = (R + r)(R - r)$. So, the area of the washer formula can be re-written as

$$\frac{2\pi(R+r)(R-r)}{2} = 2\pi\left(\frac{R+r}{2}\right)dr$$, where $dr = R - r$, which is the height or

thickness of the cylindrical shell. But $f(x) = \dfrac{R+r}{2}$, which is the average radius of the cylindrical shell. So, we have the circumference of the shell as $C = 2\pi f(x)$. The height of the shell is $h(x)$. The lateral surface area of each shell is $C*h(x)$ or $2\pi f(x)h(x)$ If we add up all of the lateral surface areas of the cylindrical shells and multiply it by their thickness dx, then we can express the volume of the generated solid rotated about the x-axis as

$$V = \int_a^b 2\pi f(x)h(x)dx.$$

Likewise, if we have a cylindrical shell that has an average radius $f(y)$ and a height $h(y)$ that is bounded between $y = c$ and $y = d$, then we can express the volume of the generated solid about the y-axis as

$$V = \int_c^d 2\pi f(y)h(y)dy.$$

Note that when we use the shell method, we slice parallel to the axis of rotation.

See the figures below that illustrate the shell method.

Volume of solid rotated about x-axis

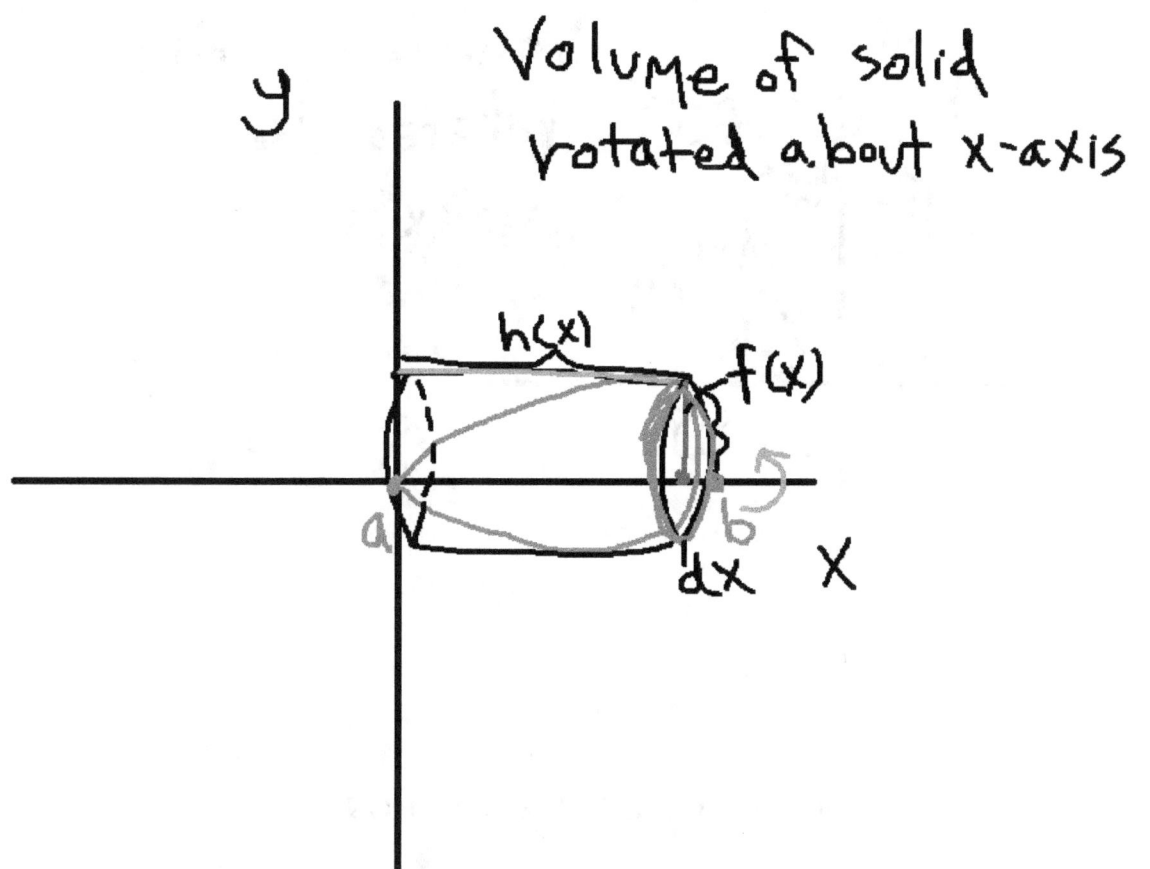

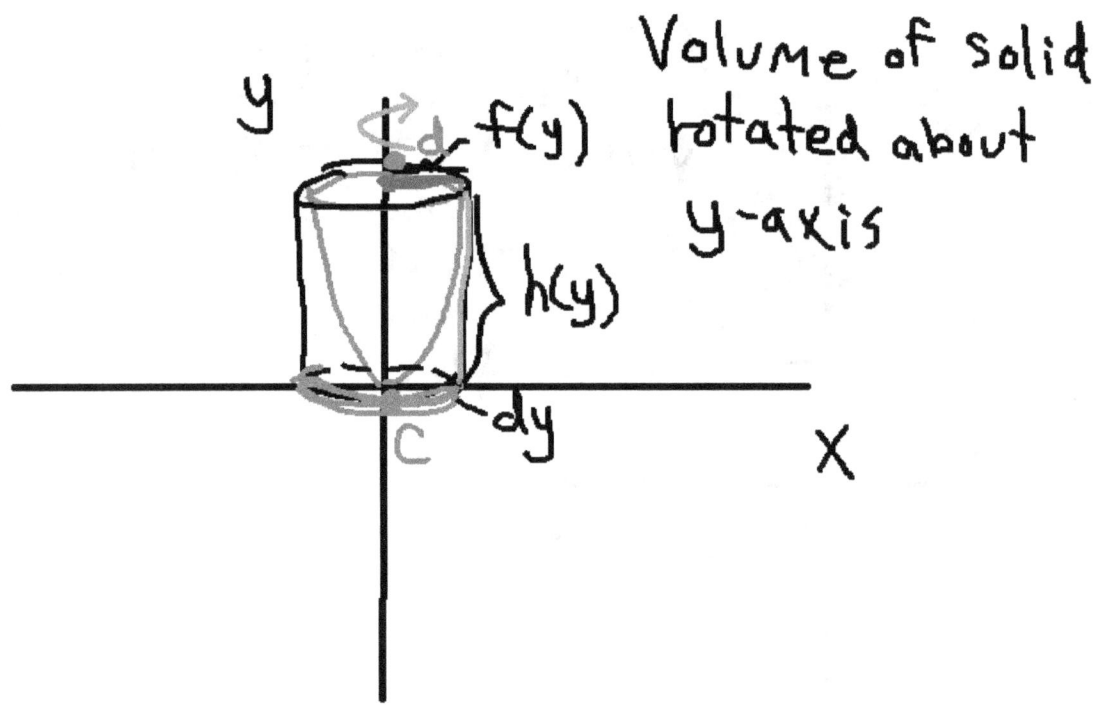

Volume of solid rotated about y-axis

AVERAGE VALUE OF A FUNCTION

Suppose we want to go outside once a day and measure the temperature with a thermometer. For a full week, we have 7 temperature readings in our data set. To find the average temperature of all the readings, we simply add them up and divide by the number of temperature readings. But what if we want to find the average temperature when we can gather an infinitely large amount of temperature readings? We can do this with calculus by computing the average value of a function $f(x)$.

AVERAGE VALUE OF A FUNCTION: Let $f(x)$ be a continuous on $[a, b]$ and differentiable. Then the average value of $f(x)$ is defined as

$$f_{avg} = \frac{1}{b-a} \int_a^b f(x)dx.$$

The proof of the average value of a function will be left as an exercise for the student.

THE MEAN VALUE THEOREM FOR INTEGRALS

In the calculus 1 review, we learned about the mean value theorem in terms of differentiation. There is also a mean value theorem in terms of integration, which is a direct consequence of the former.

THE MEAN VALUE THEOREM FOR INTEGRALS: If $f(x)$ is continuous on $[a, b]$, then there exists a c in $[a, b]$ such that

$$\int_a^b f(x)dx = f(c)(b-a).$$

PROOF:

Let $f(x)$ be continuous on $[a, b]$ and define a new function

$F(x) = \int_a^x f(t)dt.$ From the Mean Value Theorem, there exists a c in (a, b)

such that $\dfrac{f(b) - f(a)}{b - a} = f'(c).$

Now $F(b) = \int_a^b f(t)dt$ and $F(a) = \int_a^a f(t)dt = 0.$

From the Fundamental Theorem of Calculus Part 1, we have that

$F'(c) = f(c).$

Then $\dfrac{F(b) - F(a)}{b - a} = \dfrac{\int_a^b f(t)dt - 0}{b - a} = \dfrac{\int_a^b f(t)dt}{b - a} = f(c).$

So, $\int_a^b f(t)dt = f(c)(b-a).$

Making a change of variables, we have $t = x$ and $dt = dx$.

Thus, $\int_a^b f(x)dx = f(c)(b-a)$ as required. *

All the mean value theorem of integrals is saying is that we find a number c between a and b so that the area under the rectangle with width $b - a$ and height $f(c)$ is the same as the area under $f(x)$ between a and b.

In the next review we will investigate some analytical methods of computing integrals such as integration by parts, powers of trigonometric integrals, trigonometric substitution, partial fractions, etc. We will also observe how to compute improper integrals and numerical methods of integration such as the Trapezoidal Rule and Simpson's Rule to estimate definite integrals.

COMPUTATIONAL EXERCISES #16

1. Find the area bounded by the given curves.

   a) $y = x + 2, y = x^2$

   b) $x = y^2 - 6y, x = -4 - y^2$

2. Evaluate the integral. $\int_{-4}^{4} | \sqrt{4 - x} | \, dx$

3. Find the volume of the solid that is rotated about the given axis and bounded by the following curves.

   a) $y = x^{1/2}, x = 0, x = 4$; about the x-axis

   b) $y = x^2, y = 9$, about the y-axis

4. Find the volume of the solid bounded by $y = x^{1/2}, y = x$, rotated about the line $y = -1$.

5. Use the Shell Method to find the volume of the solid bounded by $x = y^{1/3}, y = 8, x = 0$ rotated about the x-axis.

6. Find the average value of the function on the given interval.

   a) $f(x) = x^4$ on $[0, 3]$

   b) $f(x) = xe^{x^2}$ on $[-2, 2]$

   c) $f(x) = \dfrac{x - 1}{x}$ on $[1, 4]$

7. The temperature in degrees Fahrenheit of a cup of hot cocoa is given by the function $T(t) = 25 + 60e^{-t/70}$, where t is the time in minutes. What is the average temperature of the hot cocoa during the first 10 minutes?

8. Find the positive value of a so that the average value of $f(x) = x^2 - 2x + 1$ on $[0, a]$ is 1.

9. Find a value c in radians so that the average value of $f(x) = \cos 2x$ is $-1/\pi$. Round your answer to three decimal places.

10. Find the average velocity of a moving vehicle on $[t_1, t_2]$. Is the answer you get related to the average velocity of a moving vehicle between t_1 and t_2? Explain.

CONCEPTUAL EXERCISES #16

1. True or False: If $f(x)$ and $g(x)$ are both continuous on $[a, b]$ and differentiable, and $f(x) > g(x)$ for all x in $[a, b]$, then
$$\int_a^b [f(x) - g(x)]dx > 0.$$

2. True or False: The disk method uses a cross-sectional area of $\pi[f(x)]^2$ or $\pi[f(y)]^2$, where $f(x)$ and $f(y)$ are the designated radii.

3. True or False: The washer method has a cross-sectional area of $\pi[r^2 - R^2]$, where r = inner radius and R = outer radius.

4. True or False: The volume of a solid with using the shell can be described geometrically as

Volume = Circumference of Cylinder x Height of Cylinder x Thickness of Cylinder.

5. True or False: If $f(x)$ is odd, then $\dfrac{1}{b-a}\int_a^b f(x)dx = \dfrac{1}{a-b}\int_b^a f(x)dx$.

6. True or False: If $f(x) = x^n$ and n is an even natural number, then the average value of $f(x)$ on $[-b, b]$ is in the form cb^p, where c is a rational number such that $0 < c < 1$ and p is an even natural number.

7. If the displacement $x(t) = \dfrac{1}{2} g t^2$ and $v = gT$, where $T = t_1 + t_2$, then

   show that the average velocity $v_{avg} = \dfrac{\Delta x}{\Delta t} = \dfrac{1}{2} v$.

8. Let $f(x)$ be continuous on $[a, b]$ and suppose $f^1(x)$ exists.

   a) If $A = \displaystyle\int_a^b f(x)dx$, then prove that there exists a c in $[a, b]$ such

   that $c = f^{-1}\left(\dfrac{A}{b-a}\right)$.

   b) Give an example that demonstrates this statement.

9. Prove the Average Value of a Function Theorem.

   Hint: Divide the interval $[a, b]$ into n number of intervals with a

   width of $\Delta x = \dfrac{b-a}{n}$ and consider the average of function values

   $f(x_1), f(x_2), \ldots, f(x_n)$.

10. Prove or Disprove. There does not exist non-zero, continuous and

    differentiable functions $f(x)$ and $g(x)$ such that $\displaystyle\int_a^b \dfrac{f(x)}{g(x)}dx = \dfrac{f(c)}{g(c)}$,

    where c is a number in $[a, b]$.

11. If n is a natural number and c is in $[a, b]$, then show that

    $$\left[\int_a^b f(x)dx\right]^n = [f(c)(b-a)]^n.$$

REVIEW #17

CALCULUS 2 PART 2

In the second part of the calculus 2 review, we are going to look at some important techniques of integration that can help us evaluate integrals that require something more than the regular rules of integration or u-substitution. We will also review improper integrals, which are integrals where either the interval is infinite or the integrand has an infinite discontinuity. And finally, we will wrap up the review with some numerical methods of integration such as Trapezoidal Rule and Simpson's Rule.

INTEGRATION BY PARTS

In calculus 1, we learned some important rules of differentiation that allowed us to calculate the derivative of various types of functions. Suppose that we want to compute $\int f(x)g(x)dx$, where $f(x)$ and $g(x)$ are two unrelated, continuous and differentiable functions on an interval I. How do we do this? We cannot use u-substitution since there is no chain or composition of functions. So, where do we go from here? The answer is to go back to the rules of differentiation with the product rule.

Recall that the product rule says

$$\frac{d}{dx}[f(x)g(x)] = f(x)g'(x) + f'(x)g(x).$$

Now if we integrate both sides of the equation with respect to x, we have the following.

$$f(x)g(x) = \int f(x)g'(x)dx + \int f'(x)g(x)dx.$$

This implies that $\int f(x)g'(x)dx = f(x)g(x) - \int f'(x)g(x)dx.$

A simpler way to write this is to let $u = f(x)$ and $v = g(x)$. Then we have $du = f'(x)\,dx$ and $dv = g'(x)\,dx$. Thus, we have

$$\int u\,dv = uv - \int v\,du.$$

This method of integration is called integration by parts.

We can do integration by parts with definite integrals, but we need to include the limits of integration.

$$\int_a^b u\,dv = uv\,|_a^b - \int_a^b v\,du$$

Now, let us look at some important integrals that involve limits of integration.

1. $\int xe^x dx = xe^x - x + C$

PROOF:

Let $u = x$ and $dv = e^x\,dx$. Then $du = dx$ and $v = e^x$. Substituting these into our integration by parts formula, we have

$$\int xe^x dx = xe^x - \int e^x dx = xe^x - e^x + C. \ *$$

Recall that $\int e^x dx = e^x + C$.

2. $\int \ln x\,dx = x\ln x - x + C$

PROOF:

Let $u = \ln x$ and $dv = dx$. Then $du = dx/x$ and $v = x$. Substituting these into our integration by parts formula, we have

$$\int \ln x\,dx = x\ln x - \int \frac{x\,dx}{x} = x\ln x - \int dx = x\ln x - x + C. \ *$$

Here is an unusual case of integration by parts, but is very effective for exponential and trigonometric functions.

3. $\int e^x \sin x\,dx = \frac{1}{2}e^x(\sin x - \cos x) + C$

PROOF:

Let $u = e^x$ and $dv = \sin x\,dx$. Then $du = e^x\,dx$ and $v = -\cos x$. Substituting these into our integration by parts formula, we have

$$\int e^x \sin x dx = -e^x \cos x + \int e^x \cos x dx.$$

We let $u = e^x$ and $dv = \cos x\, dx$. Then $du = e^x\, dx$ and $v = \sin x$. It follows that

$$\int e^x \sin x dx = -e^x \cos x + e^x \sin x - \int e^x \sin x dx.$$

If we add $\int e^x \sin x dx$ to both sides, we have

$$2\int e^x \sin x dx = e^x (\sin x - \cos x) + C.$$

Dividing by 2 to both sides gives us

$$\int e^x \sin x dx = \frac{1}{2} e^x (\sin x - \cos x) + C. \; *$$

Note that we had to use integration by parts multiple times, and then we end up the same integrand on the right hand side as we did on the left hand side. Then we finished the integral by dividing by the constant to both sides.

When doing integration by parts, it is sometimes tricky to know what to pick for u and dv. It takes some practice and trial and error to master this technique of integration.

TRIGONOMETRIC INTEGRALS

Suppose we have an integral involving the product of two different trigonometric functions with the same or different powers. Both functions may have even powers while one may have odd powers and the other has even powers. Let us consider some different forms and cases, so we can discover some general rules to help us compute such integrals.

The first form involves various powers of $\sin x$ and $\cos x$.

FORM 1: $\int \sin^n x \cos^m x dx$

CASE 1: n is odd

Let $n = 2k + 1$, where k is an integer. Then

$$\int \sin^{2k+1} x \cos^m x dx = \int \sin^{2k} x \sin x \cos^m x dx = \int (1 - \cos^2 x)^k \cos^m x \sin x dx.$$

We have a power of $\sin x$ left over.

So, we can let $u = \cos x$ and $du = \sin x\ dx$.

CASE 2: m is odd

This is similar to case 1. However, we have a power of $\cos x$ leftover instead of $\sin x$.

So, we can let $u = \sin x$ and $du = \cos x\ dx$.

CASE 3: Both m and n are even.

Let $n = 2k$, where k is an integer. And let $m = 2l$, where l is an integer. Then $\int \sin^{2k} x \cos^{2l} x dx$. Now we are a little stuck since there is no power of $\sin x$ or $\cos x$ that remains. The best way to counteract this is to utilize the half angle identities.

HALF ANGLE IDENTITIES

$$\sin^2 x = \frac{1}{2}(1 - \cos 2x)$$

$$\cos^2 x = \frac{1}{2}(1 + \cos 2x)$$

The second form involves various powers of $\sec x$ and $\tan x$.

FORM 2: $\int \sec^n x \tan^m x dx$

CASE 1: Both m and n are odd.

Let $n = 2k + 1$, where k is an integer. And let $m = 2l + 1$, where l is an integer. Then $\int \sec^{2k+1} x \tan^{2l+1} x dx =$

$\int \sec^{2k} \sec x \tan\ x \tan^{2l} x dx = \int \sec^{2k} x(\sec^2 x - 1)^l \sec x \tan x dx.$

Thus, we can let $u = \sec x$ and $du = \sec x \tan x\ dx$.

CASE 2: Both m and n are even.

For this case, we will have a $\sec^2 x$ left over. So, we can let $u = \tan x$ and $du = \sec^2 x \, dx$.

The form $\int \csc^n x \cot^m x \, dx$ is similar to the previous form with $\sec x$ and $\tan x$, so we will omit the cases here.

OTHER IMPORTANT TRIGONOMETRIC INTEGRALS

If we want to evaluate other trigonometric integrals such as

$\int \sin nx \sin mx \, dx$, $\int \cos nx \cos mx \, dx$, and $\int \sin mx \cos nx \, dx$, then we need to use the following identities.

$$\sin x \sin y = \frac{1}{2}[\cos(x-y) - \cos(x+y)]$$

$$\cos x \cos y = \frac{1}{2}[\cos(x-y) + \cos(x+y)]$$

$$\sin x \cos y = \frac{1}{2}[\sin(x-y) + \sin(x+y)]$$

TRIGONOMETRIC SUBSTITUTION

Suppose we want to evaluate integrals such as $\int \frac{dx}{\sqrt{4-x^2}}$, $\int \frac{x^3 dx}{9+x^2}$,

or $\int \frac{dx}{x\sqrt{x^2-25}}$. How do we evaluate these integrals? Like in calculus 1 with u-substitution, there is another substitution we can use called a trigonometric substitution. There are three types of trigonometric substitutions we can use for three different cases.

CASE 1: Integrand contains $a^2 - x^2$. We let $x = a\sin\theta$. Then we have $a^2 - a^2\sin^2\theta = a^2(1 - \sin^2\theta) = a^2\cos^2\theta$. And $dx = a\cos\theta\, d\theta$.

CASE 2: Integrand contains $a^2 + x^2$. We let $x = a\tan\theta$. Then we have $a^2 + a^2\tan^2\theta = a^2(1 + \tan^2\theta) = a^2\sec^2\theta$. And $dx = a\sec^2\theta\, d\theta$.

CASE 3: Integrand contains $x^2 - a^2$. We let $x = a\sec\theta$. Then we have $a^2\sec^2\theta - a^2 = a^2(\sec^2\theta - 1) = a^2\tan^2\theta$. And $dx = a\sec\theta\tan\theta\, d\theta$.

It is important to note that once we evaluate the integral in terms of angle θ, we need to rewrite our answer in terms of x.

So, for case 1 we have $\theta = \sin^{-1}(x / a)$. For case 2, we have $\theta = \tan^{-1}(x / a)$. And for case 3 we have $\theta = \sec^{-1}(x / a)$.

Here is an integral that requires trigonometric substitution.

$$\int \frac{dx}{\sqrt{a^2 - x^2}} = \sin^{-1}\left(\frac{x}{a}\right) + C$$

PROOF:

Let $x = a\sin\theta$. Then $dx = a\cos\theta\, d\theta$. And $a^2 - x^2 = a^2 - a^2\sin^2\theta = a^2(1 - \sin^2\theta) = a^2\cos^2\theta$. So, we have

$$\int \frac{dx}{\sqrt{a^2 - x^2}} = \int \frac{a\cos\theta\, d\theta}{\sqrt{a^2 \cos^2\theta}} = \int d\theta = \theta + C.$$

But $\theta = \sin^{-1}(x / a)$, so $\displaystyle\int \frac{dx}{\sqrt{a^2 - x^2}} = \sin^{-1}\left(\frac{x}{a}\right) + C.$ *

INTEGRATION BY PARTIAL FRACTIONS

When we want to evaluate an integral in the form $\displaystyle\int \frac{f(x)dx}{g(x)}$, where $g(x) \neq 0$ and u-substitution and trigonometric substitutions are not an option, then we have to apply the method of partial fractions. The student needs to review the 4 cases presented in Review #5 to evaluate integrals by means of partial fractions. Keep in mind that both $f(x)$ and $g(x)$ are polynomial functions, which is a requirement to use partial fractions.

Let us work through a couple of problems to see how integration by partial fractions work.

$$\int \frac{x+2}{x^2+3x-4}dx$$

We can factor the denominator as $(x+4)(x-1)$. So, we have

$$\frac{x+2}{x^2+3x-4} = \frac{x+2}{(x+4)(x-1)} = \frac{A}{x+4} + \frac{B}{x-1}.$$

Our goal is to find the constants A and B. Multiplying $(x+4)(x-1)$ to both sides of the equation, we get

$$x+2 = A(x-1) + B(x+4).$$

Letting $x = -4$, we have $-2 = -5A$ or $A = 2/5$.

Letting $x = 1$, we have $3 = 5B$ or $B = 3/5$.

Our integral becomes

$$\int \frac{x+2}{x^2+3x-4}dx = \int \frac{2}{5(x+4)}dx + \int \frac{3}{5(x-1)}dx = \frac{2}{5}\ln|x+4| + \frac{3}{5}\ln|x-1| + C.$$

Here is a more challenging (but not too difficult) integral by partial fractions.

$$\int \frac{x^2-4}{x^2(x^2+4)}dx$$

The denominator is already in factored form. So, we rewrite the rational expression as

$$\frac{x^2-4}{x^2(x^2+4)} = \frac{A}{x} + \frac{B}{x^2} + \frac{Cx+D}{x^2+4}.$$ Multiplying $x^2(x^2+4)$ to both sides of the equation we obtain

$$x^2-4 = Ax(x^2+4) + B(x^2+4) + x^2(Cx+D).$$

Expanding the right hand side and combining like terms we have

$$x^2-4 = (A+C)x^3 + (B+D)x^2 + 4Ax + 4B.$$

Equating the coefficients from terms on both sides we get the following equations.

$$A + C = 0$$
$$B + D = 1$$
$$4A = 0$$
$$4B = -4$$

Solving these equations, we get $A = 0$, $B = -1$, $C = 0$, and $D = 2$.

Thus, our integral becomes the following.

$$\int \frac{x^2 - 4}{x^2(x^2 + 4)} dx = \int \frac{-1}{x^2} dx + \int \frac{2}{x^2 + 4} dx$$

The first integral can be easily evaluated by the power rule, which is $1/x$. The second integral requires trig substitution.

If we let $x = 2\tan\theta$, then $dx = 2\sec^2\theta\, d\theta$. Then we have

$$\int \frac{2}{x^2 + 4} dx = \int \frac{4\sec^2\theta\, d\theta}{4\sec^2\theta} = \int d\theta = \theta + C.$$

But $\theta = \tan^{-1}(x/2)$. So, our final answer is

$$\int \frac{x^2 - 4}{x^2(x^2 + 4)} dx = \frac{1}{x} + \tan^{-1}\left(\frac{x}{2}\right) + C.$$

It is important to note that when using partial fractions we need to use the appropriate decomposition of fractions. We need to make sure the denominator is factored completely, which will help us get the right setup and make the integration process a lot smoother.

IMPROPER INTEGRALS

So far, we are familiar with how to compute $\int_a^b f(x)dx$ on a finite interval $[a, b]$. But some integrals have infinite intervals such as $(-\infty, \infty)$, $[0, \infty)$, etc. And some integrals may have infinite discontinuities as well. These types of integrals are called improper integrals. Now there are two types of improper integrals. One type is an improper integral with an infinite interval. The other type is an improper integral with an infinite discontinuity. Below are the following definitions of the improper integral.

IMPROPER INTEGRAL WITH INFINITE INTERVAL:

a) If $\int_a^t f(x)dx$ exists for all $t \geq a$, then $\int_a^\infty f(x)dx = \lim_{t \to \infty} \int_a^t f(x)dx$,

   provided that the limit exists.

b) If $\int_{-\infty}^b f(x)dx$ exists for all $t \leq b$, then $\int_{-\infty}^b f(x)dx = \lim_{t \to -\infty} \int_t^b f(x)dx$,

   provided that the limit exists.

   If the limit exists, then we say the improper integral is convergent or converges. This means the area under the curve of $f(x)$ is finite. If the limit does not exist, then we say the improper integral is divergent or diverges. This means the area under the curve of $f(x)$ is infinite.

c) Suppose that both $\int_{-\infty}^a f(x)dx$ and $\int_a^\infty f(x)dx$ are convergent. Then

$$\int_{-\infty}^\infty f(x)dx = \int_{-\infty}^a f(x)dx + \int_a^\infty f(x)dx$$

   This statement is true for any real number a.

IMPROPER INTEGRAL WITH INFINITE DISCONTINUITY:

a) Let $f(x)$ be continuous on $[a, b)$ and discontinuous at $x = b$. Then

   $\int_a^b f(x)dx = \lim_{t \to b^-} \int_a^t f(x)dx$, provided that the limit exists.

b) Let $f(x)$ be continuous on $(a, b]$ and discontinuous at $x = a$. Then

   $\int_a^b f(x)dx = \lim_{t \to a^+} \int_t^b f(x)dx$, provided that the limit exists.

c) If $f(x)$ has a discontinuity at $x = c$ in (a, b) and both $\int_a^c f(x)dx$ and

$\int_c^b f(x)dx$ are convergent, then $\int_a^b f(x)dx = \int_a^c f(x)dx + \int_c^b f(x)dx$.

COMPARISON TEST FOR IMPROPER INTEGRALS

The comparison test for improper integrals is very useful when we cannot find the exact value of an integral.

COMPARISON THEOREM: Let $f(x)$ and $g(x)$ be continuous functions with $f(x) \geq g(x) \geq 0$ for all $x \geq a$.

a) If $\int_a^\infty f(x)dx$ converges, then $\int_a^\infty g(x)dx$ converges.

b) If $\int_a^\infty g(x)dx$ diverges, then $\int_a^\infty f(x)dx$ diverges.

This theorem is visually intuitive, so we can see why it is true.
In part a), $f(x)$ is larger than $g(x)$. So, if $f(x)$ has a finite area, then $g(x)$ must also have a finite area. On the other hand, part b) says that if $g(x)$ has infinite area, then $f(x)$ must also have infinite area.

TRAPEZOIDAL RULE

The first method of numerical integration we will cover in this review is the trapezoidal rule. The trapezoidal rule tells us to divide the area under the curve into individual trapezoids and add up their individual areas. While the trapezoidal rule does not give an exact answer for the area under the curve, it does provide a great method of approximating the area under the curve for functions that are quite difficult to integrate analytically.

Suppose we want to approximate $\int_a^b f(x)dx$ by drawing individual

trapezoids of height $\Delta x = \dfrac{b-a}{n}$ and bases $f(x_n), f(x_{n+1})$. Then for each

of the endpoints $x_0, x_1, x_2, \ldots, x_n$ that make up the sub-intervals, we can approximate the area as follows.

$$\int_a^b f(x)dx \approx T_n = \frac{\Delta x}{2}\left[f(x_o) + 2f(x_1) + 2f(x_2) + \ldots + 2f(x_{n-1}) + f(x_n)\right]$$

Notice the pattern in the formula with 1, 2, 2, …, 2, 1 is easy to remember since the first and last function values have a coefficient of 1 and all of the function values sandwiched between have a coefficient of 2. This is based on the fact that the larger base of one trapezoid is shared with the smaller base of its adjacent trapezoid. So, the function value is counted more than once. Although we did not derive the trapezoidal rule, we will leave it as an exercise for the student.

ERROR OF THE TRAPEZOIDAL RULE

The error that results in numerical integration is the amount that needs to be added to make the approximation exact. Let us see how we can compute the error of the trapezoidal rule.

ERROR: Suppose $|f''(x)| \leq T$ for $a \leq x \leq b$. If T_{error} is the error

resulted from the trapezoidal rule, then $|T_{error}| \leq \dfrac{T(b-a)^3}{12n^2}$.

Notice that the error depends on the second derivative of $f(x)$, which tells us how the graph is curved.

SIMPSON'S RULE

The second method of numerical integration is more rigorous to derive, but its formula gives us a more precise approximation of the area than the trapezoidal rule. This method is known as Simpson's Rule, which in named in honor of the 18th century English mathematician Thomas Simpson. Unlike the trapezoidal rule, which involved drawing line segments to create trapezoids under the graph of *f*(*x*), Simpson's Rule allows us to draw parabolas to estimate the area. The derivation of Simpson's Rule is omitted from the text.

$$\int_a^b f(x)dx \approx S_n = \frac{\Delta x}{3}\left[f(x_o) + 4f(x_1) + 2f(x_2) + 4f(x_3) + \dots + 2f(x_{n-2}) + 4f(x_{n-1}) + f(x_n)\right]$$

Recall that $\Delta x = \dfrac{b-a}{n}$, and *n* is an even natural number.

The pattern of the coefficients are 1, 4, 2, 4, …, 2, 4, 1. This pattern is easy to remember since the first and last function values have a coefficient of 1 while the coefficients of the function values sandwiched between oscillate from 4 to 2.

ERROR OF SIMPSON'S RULE

ERROR: Suppose that $|f^{(4)}(x)| \le S$ for $a \le x \le b$. If S_{error} represents the error resulted from Simpson's Rule, then $|S_{error}| \le \dfrac{S(b-a)^5}{180n^4}$.

Notice that the error of Simpson's Rule depends on a fourth derivative of *f*(*x*), which is a higher ordered derivative required than the error of the trapezoidal rule.

Thomas Simpson was a 18th century mathematician whose main occupation was a weaver. Simpson taught himself mathematics and became a very successful mathematician in his day. He became a schoolmaster at the age of 15 and taught mathematics for several years. Even though scholars believe that Johnannes Kepler discovered the rule, it was Simpson who popularized it in his textbook called *A Treatsie of Fluxions*. Along with Newton and Leibniz, Simpson was one of the few elite mathematicians to master the rigorous concepts in infinitesimal calculus.

The next review of calculus 2 will cover a short array of applications of integration such as arc length, work, and probability.

COMPUTATIONAL EXERCISES #17

1. Use integration by parts to evaluate each of the following integrals.

   a) $\int x^2 \sin x dx$

   b) $\int_1^3 x \ln x dx$

   c) $\int \tan^{-1} x dx$

   d) $\int_0^\pi e^x \cos x dx$

2. Evaluate each of the following trigonometric integrals.

   a) $\int \sin^3 x \cos^4 x dx$

   b) $\int \tan^2 x \sec^6 x dx$

   c) $\int_{\frac{\pi}{4}}^{\frac{\pi}{3}} \sin^2 x dx$

   d) $\int_{-\frac{\pi}{2}}^{\frac{\pi}{2}} \cos 3x \sin 5x dx$

3. Use trigonometric substitution to evaluate each of the following integrals.

   a) $\int \dfrac{dx}{x\sqrt{9 - x^2}}$

   b) $\int \dfrac{x^3}{36 + x^2} dx$

c) $\displaystyle\int_{3}^{4}\frac{dx}{\sqrt{x^2-1}}$

d) $\displaystyle\int_{-8}^{0}\frac{dx}{x^2+64}$

4. Use the method of partial fractions to evaluate each of the following integrals.

a) $\displaystyle\int\frac{x^2+3x+1}{x^3+2x^2}dx$

b) $\displaystyle\int\frac{6-x}{36x^2+x^4}dx$

c) $\displaystyle\int_{0}^{1}\frac{x+4}{x^2-4}dx$

d) $\displaystyle\int_{1}^{2}\frac{dx}{x(x^2+1)}$

5. Evaluate each of the following improper integrals. State whether the integral converges or diverges.

a) $\displaystyle\int_{1}^{\infty}\frac{1}{x^2}dx$

b) $\displaystyle\int_{-\infty}^{2}\frac{1}{2-x}dx$

c) $\displaystyle\int_{-\infty}^{\infty}\frac{3dx}{x^2+9}$

d) $\displaystyle\int_{0}^{4}\frac{1}{\sqrt{4-x}}dx$

6. Use the trapezoidal rule to approximate $\int_1^3 \frac{e^x}{x} dx$ for $n = 4$. Find the trapezoidal error to four decimal places.

7. Use Simpson's rule to approximate $\int_0^6 (x^4 - x^2 + 1)dx$. How does this compare to the actual area under the curve? (Relative Error). Compute Simpson's error to four decimal places.

CONCEPTUAL EXERCISES #17

1. True or False: If $f(x)$ and $g(x)$ are two, distinct functions, then $\int f(x)g'(x)dx = f(x)g(x) - \int f'(x)g(x)dx$.

2. True or False: For the integral $\int \sin^3 x dx$, it is best to use the substitution $u = \cos^2 x$.

3. True or False: The substitution $x = 2\tan\theta$ can be used to evaluate the integral $\int \frac{dx}{4 + x^2}$.

4. True or False: $\int_a^\infty x^{-n} dx$ is convergent for all natural numbers $n \geq 2$.

5. True or False: If $g(x) \geq f(x)$ and $\int_a^\infty g(x)dx$ diverges, then $\int_a^\infty f(x)dx$ also diverges.

6. True or False: Simpson's Rule provides a greater approximation of the area under the curve than the Trapezoidal Rule.

7. True or False: Simpson's Rule can be used for any natural number n.

8. Show that $\int e^{ax}\sin(ax)dx = \dfrac{e^{ax}(\sin ax - \cos ax)}{2a} + C.$

9. Prove that $\int [f''(t)g(t) - f(t)g''(t)]dt = f'(t)g(t) - f(t)g'(t).$

   Hint: Use the integral $\int f''(t)g(t)dt$ and apply integration by parts.

10. Show that $\displaystyle\int_{-\infty}^{\infty} \dfrac{2}{4 + x^2}dx = \pi.$

11. Prove that $\displaystyle\int_{0}^{\infty} e^{-x^2} dx$ converges.

12. Prove or disprove. $\displaystyle\int \dfrac{a}{x^2 + ax}dx = \ln[x(x + a)] + C$

13. Use the Trapezoidal Rule to compute the maximum error of $\displaystyle\int_{1}^{2} \dfrac{1}{x}dx$ in terms of n. What is the minimum number of trapezoids needed if $|T_{error}| < 0.01$?

14. Use Simpson's Rule to compute the maximum error of $\displaystyle\int_{0}^{1} xe^x dx$. What is the minimum number for n if $|S_{error}| < 0.01$?

REVIEW #18

CALCULUS 2 PART 3

The part 3 review of calculus will cover some important applications such as arc length and work. The first application deals with finding the total distance along the arc between two points a and b on a given curve. The second application is a topic used in physics to calculate the work of a variable force acted on an object from point a to point b.

Suppose we have a continuous and differentiable function $f(x)$ on $[a, b]$, and we are interested in finding the length of an arc from $x = a$ to $x = b$. This presents a challenging problem if we were to rely strictly on algebra to find a solution. However, we can divide the arc into tiny line segments and then add up all of the lengths of the line segments together.

Now suppose that we have points $P_i = (x_i, y_i)$ and $P_{i+1} = (x_{i+1}, y_{i+1})$ that lie on the curve between $x = a$ and $x = b$.

We can find the distance of the small line segment P_iP_{i+1}. Then $|P_iP_{i+1}| =$
$$\sqrt{(x_{i+1} - x_i)^2 + (y_{i+1} - y_i)^2} = \sqrt{(\Delta x)^2 + (\Delta y'_i)^2},$$ where $\Delta y'_i = y_{i+1} - y_i$.

By the Mean Value Theorem, there exists an x_i^* in $[x_i, x_{i+1}]$ such that

$$f(x_{i+1}) - f(x_i) = f'(x_i^*)(x_{i+1} - x_i).$$ This implies that

$\Delta y'_i = f'(x_i^*)\Delta x$. Substituting this into the distance formula, we have

$$\sqrt{(\Delta x)^2 + [f'(x_i^*)\Delta x]^2} = \sqrt{(\Delta x)^2 + (f'(x_i^*))^2(\Delta x)^2} = \sqrt{(\Delta x)^2(1 + [f'(x_i^*)]^2)} = \sqrt{1 + [f'(x_i^*)]^2}\,\Delta x.$$

We can add up n number of these lines segments and take n to infinity to get the following Riemann sum.

$$\lim_{n \to \infty} \sum_{i=1}^{n} \sqrt{1 + [f'(x_i^*)]^2}\,\Delta x$$

This leads us to the following definition of arc length in terms of a definite integral.

ARC LENGTH: If $f'(x)$ is continuous on $[a, b]$, then the arc length of $f(x)$ from $x = a$ to $x = b$ is $L = \int_a^b \sqrt{1 + [f'(x)]^2}\,dx$.

Alternatively, we can rewrite this in Leibniz notation as

$$L = \int_a^b \sqrt{1 + \left(\frac{dy}{dx}\right)^2}\, dx.$$

On the other hand, we may have a function $x = g(y)$. Then we have the following definition.

ARC LENGTH: If $g'(y)$ is continuous on $[c, d]$, then the arc length of $g(y)$ from $y = c$ to $y = d$ is $L = \int_c^d \sqrt{1 + [g'(y)]^2}\, dy.$

The Leibniz notation is written as $L = \int_c^d \sqrt{1 + \left(\frac{dx}{dy}\right)^2}\, dy.$

Observe the figure that illustrates the arc length formula.

ARC LENGTH

Riemann Sum $\quad f(x)$

$$L = \lim_{n \to \infty} \sum_{i=1}^{n} \sqrt{1 + [f'(x_i^*)]^2}\, \Delta x$$

(x_{i+1}, y_{i+1})

(x_i, y_i)

$(x_0, y_0)\ (x_1, y_1)$

Integral Form

$$\rightarrow \quad L = \int_a^b \sqrt{1 + [f'(x)]^2}\, dx$$

THE DEFINITION OF WORK

In physics we have learned that energy is the ability to do work. How do we define work mathematically? We can start with the formula of kinetic energy, which is $K = \frac{1}{2}mv^2$, where m is the mass of the object measured in kg and v is the velocity of the object measured in m/s.

From kinematics, we have the equation that relates the square of the velocity to the distance traveled as $v^2 = 2ad$, where a is the acceleration of the object measured in m/s^2 and d is the distance the object travels in m.

Substituting this into the kinetic energy equation, we have

$K = \frac{1}{2}m(2ad) = mad$. But the force is defined as $F = ma$. Force is measured in newtons (N).

So, we have $K = (ma)d = Fd$.

Therefore, our equation for work can be defined as $W = Fd$. In other words, work is force times distance. The units for work are Newton-meters (Nm) or Joules (J).

Work can also be defined as a dot product. $W = F \bullet D = FD\cos\theta$.

The dot product version can be applied when a force such as pulling a crate with a rope at an angle respect to the ground. The work is defined as the applied force pulling on the crate with the rope times the distance the crate is pulled on the ground, times the cosine of the angle between the rope and the ground.

It is important to note that we can only use this formula if the force applied on a given object is constant. See the figure below that illustrates the definition of work.

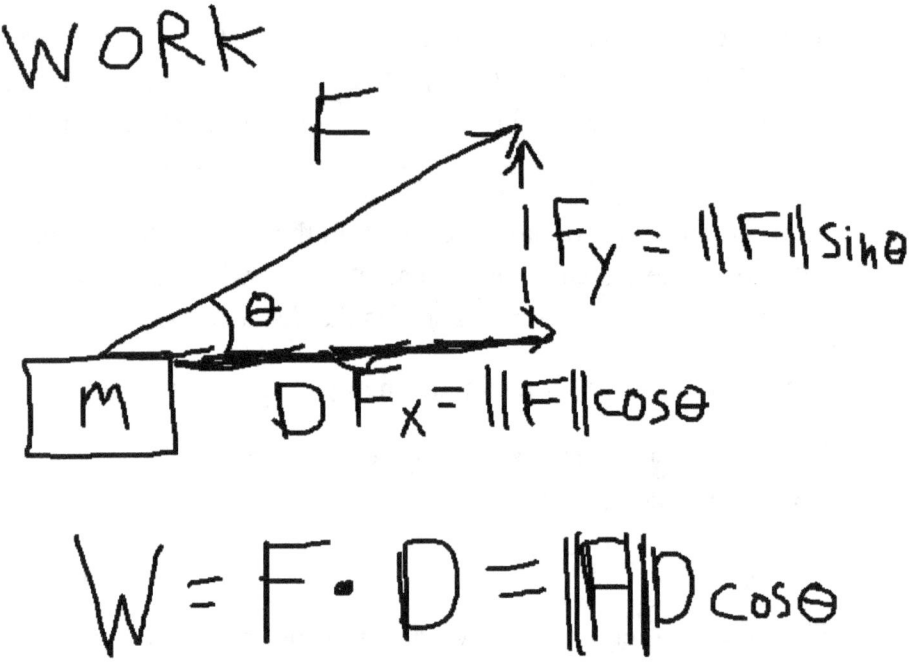

$$W = F \cdot D = \|F\| D \cos\theta$$

WORK DONE BY A VARIABLE FORCE

Sometimes the force applied on an object is not constant. At a given distance, the applied force may be stronger. And at another distance, the force may be weaker. Also, the amount of force applied on an object at a given instant varies as well. This also means that the total work done on the object for a certain distance will vary. So, how do we compute the total work done by a variable force?

We can start with the work equation earlier, which is $W = Fd$. Suppose that we let $f(x_i)$ be the force function that varies at certain points x_i. Let $\Delta x = x_{i+1} - x_i$. All we need to do is add up all the forces applied at equal, small distances and let n approach infinity.

Then the total work done from $x = x_1$ to $x = x_2$ can be defined as a Riemann sum.

$$W = \lim_{n \to \infty} \sum_{i=1}^{n} f(x_i) \Delta x.$$

This leads to the integral definition of work done on a variable force.

WORK DONE BY A VARIABLE FORCE: If $f(x)$ is continuous on $[x_1, x_2]$, then the total work done on an object by the variable force $f(x)$ is

$$W = \int_{x_1}^{x_2} f(x)\,dx.$$

When the work is done on an object such as a spring, then we use an important law from physics called Hooke's Law $f(x) = kx$. Hooke's Law says that the force exerted on the spring is proportional to the spring's displacement. The constant of proportionality k is the spring constant, which is measured in N / m. If the spring constant is small, then the spring is very loose or flexible. But if the spring constant is large, then the spring is very rigid or stiff.

COMPUTATIONAL EXERCISES #18

1. Find the length of the arc of $f(x) = 3x^2$ from $x = 1$ to $x = 4$.

2. Find the length of the arc of $g(y) = y^{1/2}$ from $y = 0$ to $y = 9$.

3. Find the arc length function $L(x) = \int_{1}^{x} \sqrt{1 + [f'(x)]^2}\,dx$ if $f(x) = \ln x$.

4. Find the amount of work done by a force of 250 N at a distance of 35 m.

5. A man pulls a crate 62 ft with a force of 117 lbs at a 30° angle respect to the ground. Find the total work done on the crate.

6. An object is moved 6 m from the origin by a variable force of $f(x) = x^2 + 4x$. Find the total work done on the object.

7. A 20 cm spring is stretched 8 cm from its natural length with a force of 50 N. Find the work needed to move the spring a distance of 7 cm.

CONCEPTUAL EXERCISES #18

1. True or False: If $f(x) = x$, then the arc length of $f(x)$ on $[0, \frac{1}{2}]$ is $\dfrac{\sqrt{2}}{2}$.

2. True or False: It is possible to compute the arc length of $f(x) = 1/x$ on $[0, 1]$.

3. True or False: The maximum work done on an object is when the angle between the force and displacement is $0°$.

4. True or False: If the spring constant is divided, then the work increases by a factor of 4.

5. True or False: The total work done on an object is zero if and only if the angle between the force and distance is $90°$.

6. If c is a real constant, then show that the length of $f(x) = c$ on $[a, b]$ is $b - a$.

7. If $x*$ represents the average point between the initial point x_1 and the final point x_2 of a moving spring and $\Delta x = x_2 - x_1$, then show that the work done on the spring is $W = kx*\Delta x$.

8. If a man pulls an object a distance x with force F, and then pulls the same object a distance of $2x$ with force F at a $30°$ angle, then what is the ratio of the first work W_1 to the second work W_2?

REVIEW #19

CALCULUS 2 PART 4

In this review, we are going to look at parametric equations/curves, the calculus of parametric equations, and area in terms of polar coordinates.

PARAMETRIC EQUATIONS/CURVES

There are some curves that do not represent a function $y = f(x)$ since it fails the vertical line test. But we can introduce a new variable t (also known as a parameter) to help us plot the curve.

PARAMETRIC CURVE: If x and y are both functions of the parameter t, i.e. $x = f(t)$ and $y = g(t)$, then we can plot each coordinate $(x, y) = (f(t), g(t))$ as t varies. This curve is known as a parametric curve.

Parametric equations not only tell us the shape of the parametric curve, but they also indicate what direction the curve is progressing as t varies. For example, the set of parametric equations $x = \cos t$ and $y = \sin t$ on $0 \leq t \leq 2\pi$ is the unit circle that starts at $(1, 0)$ and progresses in a counterclockwise fashion. On the other hand, the set of parametric equations $x = \sin t$ and $y = \cos t$ on $0 \leq t \leq 2\pi$ is the unit circle that starts at $(0, 1)$ and progresses in a clockwise fashion.

In general, the parametric curve with parametric equations $x = f(t)$ and $y = g(t)$ on the interval $a \leq t \leq b$ has an initial point $(f(a), g(a))$ and a terminal point $(f(b), g(b))$.

Now suppose that we have a function $x = g(y)$. Then we can let $y = t$ and $x = g(t)$. Notice that y takes on the parameter t and x is now a function of t, which makes it easier to plot. For example, if we have $x = y^3 - 2y^2 + 1$, then we can let $y = t$ and then $x = t^3 - 2t^2 + 1$.

ELIMINATING THE PARAMETER

Sometimes we have a set of parametric equations and we want to express it in terms of a function $y = f(x)$ with rectangular coordinates. The way we do this is to eliminate the parameter t by methods of such as substitution, elimination, etc.

For instance, if we have the parametric equations $x = t - 1$ and $y = t^{1/3}$, then we solve for t in $x = t - 1$ and substitute into $y = x^{1/3}$. Thus, $t = x + 1$ which implies that $y = (x + 1)^{1/3}$. The parameter t has been eliminated, and we have the function $f(x) = (x + 1)^{1/3}$ in terms of rectangular coordinates.

THE CALCULUS OF PARAMETRIC EQUATIONS/CURVES

Suppose that we have parametric equations $x = f(t)$ and $y = g(t)$ that describe the curve $y = F(x)$. Then substitution y and x into $y = F(x)$, we have $g(t) = F(f(t))$. Now if $f(t)$, $g(t)$, and $F(x)$ are differentiable functions, then $g'(t) = F'(f(t))f'(t) = F'(x)f'(t)$. Solving for $F'(x)$, we have $F'(x) = g'(t) / f'(t)$, provided that $f'(t) \neq 0$.

This equation tells us that we can find the slope of the tangent line of $y = F(x)$ at a point $(x, F(x))$ without eliminating the parameter t.

We can rewrite the equation for $F'(x)$ in Leibniz notation as

$$\frac{dy}{dx} = \frac{\dfrac{dy}{dt}}{\dfrac{dx}{dt}}, \text{ if } \frac{dx}{dt} \neq 0.$$

Now $\dfrac{d^2 y}{dx^2} = \dfrac{d}{dx}\left(\dfrac{dy}{dx}\right) = \dfrac{d}{dx}\left(\dfrac{\dfrac{dy}{dt}}{\dfrac{dx}{dt}}\right) = \dfrac{\dfrac{d}{dt}\left(\dfrac{dy}{dx}\right)}{\dfrac{dx}{dt}}, \text{ if } \dfrac{dx}{dt} \neq 0.$

In order to find the horizontal tangent lines, we set $\dfrac{dy}{dt} = 0$. And to find the vertical tangent lines, we set $\dfrac{dx}{dt} = 0$.

ARC LENGTH IN PARAMETRIC FORM

We reviewed how to calculate the length of an arc with $y = f(x)$ in rectangular form. We can now convert the rectangular form to parametric form. Observe the following theorem.

THEOREM: Let C be a curve described by the parametric equations $x = f(t)$ and $y = g(t)$ on $[A, B]$ and $f'(t)$ and $g'(t)$ are continuous on $[A, B]$. If C is traversed only once as t increases from A to B, then the arc length of C is

$$L = \int_A^B \sqrt{\left(\frac{dx}{dt}\right)^2 + \left(\frac{dy}{dt}\right)^2}\, dt.$$

We will not give a formal proof to this theorem, but we can derive this equation from the rectangular arc length equation

$$L = \int_a^b \sqrt{1 + [f'(x)]^2}\, dx.$$

We know that $\dfrac{dy}{dx} = \dfrac{\frac{dy}{dt}}{\frac{dx}{dt}}$, which implies that

$$\sqrt{1 + \left(\frac{dy}{dx}\right)^2} = \sqrt{1 + \left(\frac{\frac{dy}{dt}}{\frac{dx}{dt}}\right)^2} = \sqrt{1 + \frac{\left(\frac{dy}{dt}\right)^2}{\left(\frac{dx}{dt}\right)^2}} = \frac{\sqrt{\left(\frac{dx}{dt}\right)^2 + \left(\frac{dy}{dt}\right)^2}}{\frac{dx}{dt}}.$$

Since $dx = \dfrac{dx}{dt}\, dt$, and $A \leq t \leq B$ we have

$$L = \int_a^b \sqrt{1 + \left(\frac{dy}{dx}\right)^2}\, dx = \int_A^B \frac{\sqrt{\left(\frac{dx}{dt}\right)^2 + \left(\frac{dy}{dt}\right)^2}}{\frac{dx}{dt}} \frac{dx}{dt}\, dt$$

$$= \int_A^B \sqrt{\left(\frac{dx}{dt}\right)^2 + \left(\frac{dy}{dt}\right)^2}\, dt.$$

TANGENTS TO POLAR CURVES

Suppose that we have a polar equation $r = f(\theta)$ that represents a polar curve and θ is the given parameter. Then we know from polar coordinates that $x = r\cos\theta$ and $y = r\sin\theta$. This implies that $x = f(\theta)\cos\theta$ and $y = f(\theta)\sin\theta$.

The tangent can be found by

$$\frac{dy}{dx} = \frac{\frac{dy}{d\theta}}{\frac{dx}{d\theta}} = \frac{\frac{dr}{d\theta}\sin\theta + r\cos\theta}{\frac{dr}{d\theta}\cos\theta - r\sin\theta}.$$

The horizontal tangent(s) can be found by solving the equation

$$\frac{dy}{d\theta} = 0.$$

The vertical tangent(s) can be found by solving the equation

$$\frac{dx}{d\theta} = 0.$$

AREA WITH POLAR COORDINATES

We know from trigonometry that the area of a sector is given by the formula $A = \frac{1}{2}r^2\theta$. We can use this formula to find the area of within a polar curve.

Suppose we have the polar function $r = f(\theta)$, where $r > 0$ and continuous, and let R be a region bounded by the polar curve generated by r. Divide the interval $[a, b]$ into sub-intervals with endpoints $\theta_0, \theta_1, \theta_2, \ldots, \theta_n$ with equal width of $\Delta\theta$. We can choose an angle θ_i^* in $[\theta_i, \theta_{i+1}]$ so that the differential area

$$\Delta A \approx \frac{1}{2}[f(\theta_i^*)]^2 \Delta\theta.$$

Now adding up all of these differential areas and taking n to infinity gives us the Riemann sum

$$\lim_{n \to \infty} \sum_{i=1}^{n} \frac{1}{2}[f(\theta_i)]^2 \Delta\theta.$$

By definition, the total area of the polar region is an integral.

$$A = \int_a^b \frac{1}{2}[f(\theta)]^2 \, d\theta$$

AREA BETWEEN TWO POLAR FUNCTIONS

Suppose we have two polar functions $r = f(\theta)$ and $r = g(\theta)$ such that $f(\theta) \geq g(\theta) \geq 0$ and the region R is bounded by $f(\theta)$ and $g(\theta)$ on $[a, b]$. Then the area between the two polar functions is

$$A = \int_a^b \frac{1}{2}([f(\theta)]^2 - [g(\theta)]^2)d\theta.$$

To find the area between the two polar curves, we need to do four simple steps.

1. Sketch the polar equations either by hand or with a calculator/computer.

2. Find the points of intersections to determine limits of integration.

3. Set up the integral. Note: Outside Curve – Inside Curve

4. Evaluate the integral.

The next review in calculus 2 will cover sequences and series in more detail. We will be able to determine whether a sequence/series converges or diverges by applying certain theorems and tests, how to calculate the sum of a convergent series, estimating sums of a series, along with finding the Taylor and MacLaurin Series of $f(x)$ and more.

COMPUTATIONAL EXERCISES #19

1. Sketch each parametric curve and indicate the direction of the curve as t increases.

    a) $x = 1 + t, y = t^2 - 2t, [0, 4]$

    b) $x = 3\cos t, y = 4\sin t, [0, 2\pi]$

    c) $x = t^{1/2}, y = t^2, [0, 3]$

2. Eliminate the parameter t to find a rectangular equation of the curve.

    a) $x = 2t - 4, y = 4t + 3$

    b) $x = t^2 - 1, y = 9t, [0, 5]$

3. Find $\frac{dy}{dx}$ and $\frac{d^2y}{dx^2}$ for each set of the parametric equations.

    a) $x = t + e^t, y = te^t$

b) $x = t^3 + 3t^2 - 2, y = t^2 - 4t$

4. Find the points where the horizontal and vertical tangents occur of $x = \sin t$ and $y = \sin t \cos t$.

5. Find an equation for the tangent line of the curve $x = e^t, y = t^2 + 1$ at the point (1, 1).

6. Find the length of the curve. Round to three decimal places.

   a) $x = 2t^2 + 1, y = t^3 + 8$ on [0, 2]

   b) $x = e^t - e^{-t}, y = e^t$ on [0, 1]

7. Find the slope of the tangent line at the given value of θ.

   a) $r = 3\cos\theta$ at $\theta = \pi / 8$

   b) $r = 2 + \sin\theta$ at $\theta = \pi / 6$

8. Sketch the polar curve and find the area.

   a) $r = 4\sin\theta$

   b) One loop of $r = \cos3\theta$

9. Find the area inside $1 + 2\sin\theta$ and outside $r = 2$.

10. Find the points of intersection for $r = \cos2\theta$ and $r = \cos\theta$.

11.

CONCEPTUAL EXERCISES

1. True or False: If $x = f(t)$ and $y = g(t)$, and both $f'(t)$ and $g'(t)$ are continuous on [a, b], then $L = \int_a^b \sqrt{[f'(t)]^2 + [g'(t)]^2}\, dt$.

2. True or False: If $x = f(t)$ and $y = g(t)$ are differentiable, then $\dfrac{dy}{dx} = \dfrac{g'(t)}{f'(t)}$.

3. True or False: If $f(t)$ and $g(t)$ are twice differentiable and continuous on $[a, b]$, then $\dfrac{d^2 y}{dx^2} = \dfrac{\dfrac{d^2 y}{dt^2}}{\dfrac{d^2 x}{dt^2}}$.

4. True or False: If $g(\theta) \geq f(\theta) \geq 0$ on $[a, b]$, then the area between the two polar curves is $A = \dfrac{1}{2} \int\limits_{a}^{b} ([g(\theta)]^2 - [f(\theta)]^2) d\theta$.

5. True or False: If a and b are positive numbers such that $a > b$, then the set of parametric equations $x = a\cos t$ and $y = b\sin t$ represents an ellipse centered at the origin with vertices $(-a, 0)$ and $(a, 0)$.

6. If $a, b > 0$, then show that the set of parametric equations $x = at + b$ and $y = at^2$ on $[0, 1/a]$ represents a parabola in the standard form $y = Ax^2 + Bx + C$. What is the domain?

7. Verify that $x = \sin\theta$ and $y = \cos\theta$ represents the unit circle and that the circle moves in a clockwise direction.

8. Prove that the length of the curve $x = r\cos t$, $y = r\sin t$ on $[0, 2\pi]$ is the circumference of a circle.

9. If $r = \cos\theta$, then show that

   a) $\dfrac{dy}{dx} = -\cot 2\theta$ and $\dfrac{d^2 y}{dx^2} = -2\csc^3 2\theta$

   b) $\left(\dfrac{dx}{d\theta}\right)^2 + \left(\dfrac{dy}{d\theta}\right)^2 = 1$

10. If $r = \sin\theta$, then show that the equation of the tangent line at $(1/2, \pi/6)$ is $y = \sqrt{3}x + 1$.

CALCULUS 2 PART 5

THE CALCULUS OF SEQUENCES AND SERIES

We are now going to examine the calculus aspects of sequences and series. In calculus, we want to determine whether a sequence or a series is convergent or divergent. Then we apply some methods to ascertain these facts. Let us start with the concept of sequences.

A sequence a_n has a limit L, which is notated as

$$\lim_{n \to \infty} a_n = L.$$ If the limit exists, then the sequence converges to L. Otherwise, the sequence diverges to either ∞ or $-\infty$.

PRECISE DEFINITION OF SEQUENCES: A sequence a_n has the limit L if for any $\varepsilon > 0$, there exists an integer N such that if $n > N$ then $|a_n - L| < \varepsilon.$

It is important to note that the precise definition of sequences is very similar to the precise definition of a limit for a real function $f(x)$. We basically transferred this concept from real functions to sequences. This leads us to an important theorem of sequences.

THEOREM: If $\lim_{x \to \infty} f(x) = L$ and $f(n) = a_n$ for any natural number n,

then $\lim_{n \to \infty} a_n = L.$

We also have a definition for infinite limits of a sequence.

DEFINITION OF INFINITE LIMITS:

$$\lim_{n \to \infty} a_n = \infty$$ if and only if for any $M > 0$, there exists an integer N such that if $n > N$ then $a_n > M.$

We also have the limit laws for sequences.

LIMIT LAWS FOR SEQUENCES:

1. $\displaystyle\lim_{n \to \infty} (a_n \pm b_n) = \lim_{n \to \infty} a_n \pm \lim_{n \to \infty} b_n$

2. $\displaystyle\lim_{n \to \infty} c a_n = c \lim_{n \to \infty} a_n$, for any real constant c.

3. $\displaystyle\lim_{n \to \infty} a_n b_n = \lim_{n \to \infty} a_n \lim_{n \to \infty} b_n$

4. $\displaystyle\lim_{n \to \infty} (a_n / b_n) = \lim_{n \to \infty} a_n / \lim_{n \to \infty} b_n$, where $\displaystyle\lim_{n \to \infty} b_n \neq 0$

5. $\displaystyle\lim_{n \to \infty} c = c$, for any real constant c.

6. $\displaystyle\lim_{n \to \infty} a_n{}^p = \left(\lim_{n \to \infty} a_n \right)^p$, if $a_n \geq 0$ for all n.

Here is a useful theorem that can be used for the absolute value of a sequence.

THEOREM: If $\displaystyle\lim_{n \to \infty} |a_n| = 0$ then $\displaystyle\lim_{n \to \infty} a_n = 0.$

Just like functions, we can analyze the behavior of sequences. We may want to know if a sequence is either increasing or decreasing and whether it is bounded or unbounded. Here are some important definitions.

MONOTONIC SEQUENCE: A sequence a_n is increasing if $a_n < a_{n+1}$ for all $n \geq 1$ or decreasing if $a_n > a_{n+1}$ for all $n \geq 1$. The sequence a_n is monotonic if it is either increasing or decreasing.

BOUNDED SEQUENCE: A sequence a_n is bounded above if there is a real number U so that $a_n \leq U$ for all $n \geq 1$. A sequence a_n is bounded below if there is a real number L so that $L \leq a_n$ for all $n \geq 1$. If a_n is bounded above and bounded below, then a_n is called a bounded sequence.

How do the ideas of monotonic and bounded tie in with convergence? The follow theorem answers this question.

MONTONIC SEQUENCE THEOREM: If a_n is bounded and monotonic, then a_n is convergent.

In order to prove this theorem, we need a statement called the Completeness Axiom of the real numbers.

COMPLETENESS AXIOM: Let S be a non-empty subset of the real numbers R. If S has an upper bound B, i.e. $x \leq B$ for all x in S, the S has a least upper bound b.

From the Completeness Axiom, we know that b is an upper bound for S. Now if B is any other upper bound, then $b \leq B$.

The Completeness Axiom demonstrates that there are no open holes or gaps in the real number line. Thus, the real number line is complete.

We will now prove the Monotonic Sequence Theorem for the increasing case. A similar proof can be used for the decreasing case for the greatest lower bound.

PROOF: Let a_n be a monotonic and bounded sequence. Since a_n is monotonic, it is either increasing or decreasing. Let us consider the increasing case. We know that $S = \{a_n \mid n \geq 1\}$ is a non-empty subset of R and has a least upper bound L by the Completeness Axiom. Now if $\varepsilon > 0$, then $L - \varepsilon$ is not an upper bound for S since L is the least upper bound. This means that $a_N > L - \varepsilon$ for some integer N. Since a_n is increasing and for $n > N$, we know that $a_n \geq a_N$. So, $a_n > L - \varepsilon$, which implies that $L - a_n < \varepsilon$. And since $a_n \leq L$, we know that $L - a_n \geq 0$. So, we have $0 \leq L - a_n < \varepsilon$. Therefore, if $n > N$ then $|L - a_n| < \varepsilon$. So, $\lim\limits_{n \to \infty} a_n = L$ and a_n is convergent. *

Now that we have finished covering the basics on the calculus of sequences, we can move forward to the calculus of series.

SERIES

Recall that a series is defined as the sum of the terms of a given sequence. We can denote the kth partial sum of a series as

$$s_k = \sum_{n=1}^{k} a_n = a_1 + a_2 + \ldots + a_k.$$

If $\lim\limits_{k \to \infty} s_k = s$, then the series $\sum a_n$ is convergent.

That is, $\sum\limits_{n=1}^{\infty} a_n = a_1 + a_2 + \ldots = s$.

We also recall the geometric series $\sum\limits_{n=1}^{\infty} ar^{n-1} = a + ar + ar^2 + \ldots + ar^{n-1} + \ldots$

If $|r| < 1$, then the series converges. But if $|r| \geq 1$, then series diverges.

It is important to keep in mind that a convergent series has a "finite" sum while a divergent series does not have a finite sum.

Now it is time to go over some important theorems involving series.

THEOREM: If $\sum\limits_{n=1}^{\infty} a_n$ is convergent, then $\lim\limits_{n \to \infty} a_n = 0$.

PROOF: Let $s_n = \sum\limits_{n=1}^{\infty} a_n$ be convergent. Then $s_n = a_1 + a_2 + \ldots + a_n$. So, we have $a_n = s_n - s_{n-1}$. Since $\sum a_n$ is convergent, then s_n is also convergent. If $\lim\limits_{n \to \infty} s_n = s$, then it follows that $\lim\limits_{n \to \infty} s_{n-1} = s$ since we have $n-1 \to \infty$ as $n \to \infty$.

So, $\lim\limits_{n \to \infty} a_n = \lim\limits_{n \to \infty} (s_n - s_{n-1}) = s - s = 0$. *

If we take the contrapositive of this statement, then get an equivalent statement known as the Divergence Test.

DIVERGENCE TEST: If $\lim\limits_{n \to \infty} a_n \neq 0$ then $\sum\limits_{n=1}^{\infty} a_n$ is divergent or $\lim\limits_{n \to \infty} a_n$ does not exist.

The following theorems below state the sum and difference of two convergent series is also convergent and the constant times a convergent series is also convergent.

CONVERGENT SERIES THEOREMS: Suppose that $\sum a_n$ and $\sum b_n$ are both convergent series. Then we have the following statements.

a) $\displaystyle\sum_{n=1}^{\infty} ca_n = c\sum_{n=1}^{\infty} a_n$

b) $\displaystyle\sum_{n=1}^{\infty} (a_n + b_n) = \sum_{n=1}^{\infty} a_n + \sum_{n=1}^{\infty} b_n$

c) $\displaystyle\sum_{n=1}^{\infty} (a_n - b_n) = \sum_{n=1}^{\infty} a_n - \sum_{n=1}^{\infty} b_n$

We can also compare a series to another series to determine if the original series is convergent or divergent. This is known as the Comparison Test.

COMPARISON TEST: Suppose that $\sum a_n$ and $\sum b_n$ are series with positive terms.

a) If $\sum b_n$ is convergent and $a_n \le b_n$, then $\sum a_n$ is also convergent.

b) If $\sum b_n$ is divergent and $a_n \ge b_n$, then $\sum a_n$ is also divergent.

All this is telling us is that if the larger series is convergent, then the smaller series is also convergent. On the other hand, if the smaller series is divergent, then the larger series is also divergent.

A variation of the comparison test is the Limit Comparison Test.

LIMIT COMPARISON TEST: Let $\sum a_n$ and $\sum b_n$ have positive terms.

If $\displaystyle\lim_{n \to \infty} \frac{a_n}{b_n} = c$, where c > 0 and a finite number, then either both $\sum a_n$ and $\sum b_n$ converge or diverge.

PROOF: Let l and u be positive numbers so that $l < c < u$. Since $\dfrac{a_n}{b_n} \to c$ as $n \to \infty$, there is an integer N such that $l < \dfrac{a_n}{b_n} < u$ whenever $n > N$, which implies that $lb_n < a_n < ub_n$.

Now we have two cases to consider.

Case 1: If $\sum b_n$ is convergent, then $\sum ub_n$ is also convergent. By the comparison test, $\sum a_n$ is convergent.

Case 2: If $\sum b_n$ is divergent, then $\sum lb_n$ is also divergent. By the comparison test, $\sum a_n$ is divergent. *

The next test relates the sum of an infinite series to an improper integral, which is known as the Integral Test.

INTEGRAL TEST: Suppose that $f(x) > 0$ and decreasing on $[1, \infty)$ and $a_n = f(n)$. Then $\sum\limits_{n=1}^{\infty} a_n$ is convergent if and only if $\int\limits_{1}^{\infty} f(x)dx$ is convergent.

There is a particular kind of series that follows from the application of the integral test.

P-SERIES: $\sum\limits_{n-=1}^{\infty} \dfrac{1}{n^p}$ converges if $p > 1$ and diverges if $p \le 1$.

This fact of the p-series can be verified by applying the integral test.

The famous harmonic series $\sum\limits_{n=1}^{\infty} \dfrac{1}{n}$ diverges since $p = 1$. We can see why this is true. If $f(x) = \dfrac{1}{x}$, then $\int\limits_{1}^{\infty} \dfrac{1}{x}dx = \ln x \big|_1^{\infty} = \infty$. The improper integral diverges, so the harmonic series diverges.

On the other hand, the p-series $\sum\limits_{n=1}^{\infty} \dfrac{1}{n^2}$ converges since $p = 2 > 1$. Leonard Euler proved that the sum of this series is $\dfrac{\pi^2}{6}$. The proof of this sum is very difficult and requires some advanced mathematical machinery to arrive at such a result.

We have learned that some infinite series converge while others diverge. But how can we find the exact sum of a convergent series? The answer is not obvious. However, we can estimate the sum of a series by taking a closer look at the remainder (error) of the sum.

If s is the exact sum of the series and s_n is the nth partial sum of the series, then the remainder $R_n = s - s_n$. This idea leads to an important theorem.

THEOREM: Let $f(n) = a_n$ where $f(x)$ is continuous, positive, and decreasing for $x \geq n$ and suppose that $\sum a_n$ is convergent.

If $R = s - s_n$, then $\displaystyle\int_{n+1}^{\infty} f(x)dx \leq R_n \leq \int_{n}^{\infty} f(x)dx$. This implies that

$$s_n + \int_{n+1}^{\infty} f(x)dx \leq s \leq s_n + \int_{n}^{\infty} f(x)dx.$$

What this theorem is telling us is that we can find a lower and upper bound for the sum of the series based on the nth partial sum and the small and large areas under the curve of $f(x)$.

ALTERNATING SERIES

So far, we have looked at series with positive terms only. But some series contain both positive and negative terms that alternate one after another. For example, we might have a series $1 - 2 + 3 - 4 + \ldots$ Such series are called alternating series.

If we consider the alternating series $\displaystyle\sum_{n=1}^{\infty} \frac{(-1)^n}{n} = -1 + \frac{1}{2} - \frac{1}{3} + \ldots$, how do we know if the series converges? This question leads us to another series test called the Alternating Series Test.

ALTERNATING SERIES TEST: If $\displaystyle\sum_{n=1}^{\infty} (-1)^{n+1} a_n = a_1 - a_2 + a_3 - a_4 + \ldots$, where $a_n > 0$ for all $n \geq 1$, then the following conditions must satisfy for convergence.

A) $a_{n+1} \leq a_n$ for all $n \geq 1$

B) $\displaystyle\lim_{n \to \infty} a_n = 0$

The Alternating Series Test leads us to an important theorem to estimate the sum of an alternating series.

270

ALTERNATING SERIES ESTIMATION THEOREM:

If $s = \sum (-1)^{n+1} a_n$ is the sum of the alternating series that satisfies both conditions from the alternating series test, then $|R_n| = |s - s_n| \le a_{n+1}$

This theorem says that the size of the error is less than or equal to the $n + 1$ term of the series. To see why this is true, we take two consecutive partial sums, where the actual sum s lies between s_n and s_{n+1}.

So, we have $|s - s_n| \le |s_{n+1} - s_n| = a_{n+1}$.

The question is how do we know for certain that the alternating series converges since we have to account for both positive and negative terms? This question leads us to the idea of absolute and conditional convergence.

ABSOLUTE CONVERGENCE: Consider the series $\sum |a_n|$.

If $\sum |a_n|$ is convergent, then $\sum a_n$ is absolutely convergent.

CONDITIONAL CONVERGENCE: If $\sum a_n$ is convergent, but not absolutely convergent, then $\sum a_n$ is conditionally convergent.

THEOREM: If $\sum a_n$ is absolutely convergent, then $\sum a_n$ is convergent.

PROOF: We know that $|a_n| = a_n$ or $-a_n$ by the definition of absolute value.

This implies that $0 \le a_n + |a_n| \le 2|a_n|$. Since $\sum |a_n|$ converges, then it follows that $2 \sum |a_n|$ also converges. By the comparison test, we know that $\sum (a_n + |a_n|)$ is a convergent series. Now we have $\sum a_n = \sum (a_n + |a_n|) - \sum |a_n|$, which is the difference of two convergent series. The difference of two series is convergent. Thus, the series $\sum a_n$ is also convergent. *

This fact tells us that a series that has absolute convergence guarantees convergence. On the other hand, if a series is convergent, then it may or may not be absolutely convergent.

Now we examine an important test for series that looks at the ratio of the $n + 1$ term and nth term of the series. This is known as the Ratio Test.

RATIO TEST

The Ratio Test is a series test for absolute convergence. We consider three possible cases.

Suppose that $\lim_{n \to \infty} \left| \dfrac{a_{n+1}}{a_n} \right| = R.$

R is considered to be the ratio.

A) If $R < 1$, then $\sum a_n$ is absolutely convergent.

B) If $R > 1$, then $\sum a_n$ is divergent.

C) If $R = 1$, then the test is inconclusive. So, $\sum a_n$ may or may not converge.

Another test we can employ for series with radicals is called the Root Test.

ROOT TEST

Suppose that $\lim_{n \to \infty} \sqrt[n]{|a_n|} = r.$

r is considered to be the root.

A) If $r < 1$, then $\sum a_n$ is absolutely convergent.

B) If $r > 1$, then $\sum a_n$ is divergent.

C) If $r = 1$, then the test is inconclusive. So, $\sum a_n$ may or may not converge.

Another type of series we want to investigate is the power series.

POWER SERIES

A power series can be defined as $\displaystyle\sum_{n=0}^{\infty} a_n x^n = a_0 + a_1 x + a_2 x^2 + \ldots$, where the coefficients are the a_n's. Power series may converge for some or all values of x or may diverge as well.

Consider the power series $\displaystyle\sum_{n=0}^{\infty} x^n = 1 + x + x^2 + \ldots$

This power series converges if $|x| < 1$ and diverges if $|x| \geq 1$.

The general form of a power series is given by

$$\sum_{n=0}^{\infty} a_n (x-a)^n = a_0 + a_1(x-a) + a_2(x-a)^2 + \ldots, \text{ where the series is}$$

centered about $x = a$. There are three possible events that can occur with the general form of the power series, which leads us to our next theorem.

THEOREM: For any power series $\displaystyle\sum_{n=0}^{\infty} a_n (x-a)^n$,

A) If $x = a$, then the series converges.

B) The series converges for all values of x.

C) There is a $R > 0$ such that if $|x - a| < R$, then the series converges, and if $|x - a| > R$, then the series diverges.

R is called the radius of convergence. The set of values that guarantee convergence of the power series is called the interval of convergence.

Some possible intervals of convergence that may occur if $|x - a| < R$ are

$(a - R, a + R)$, $[a - R, a + R)$, $(a - R, a + R]$, and $[a - R, a + R]$.

It is important to check the endpoints of the interval convergence since they series may or may not converge for these values of x.

REPRESENTING FUNCTIONS AS A POWER SERIES

From college algebra, we know the sum of an infinite, geometric series is $S = \dfrac{a}{1-r}$, where a is the first term in the series and r is the common ratio.

Now we can find the sum to the power series $\sum_{n=0}^{\infty} x^n$ since $a = 1$ and $r = x$.

Thus, our sum is $S = \dfrac{1}{1-x}$. We can write other functions as a power

series. For example, the function $f(x) = \dfrac{1}{1+x} = \dfrac{1}{1-(-x)}$ can be written as

$$\sum_{n=0}^{\infty} (-x)^n = \sum_{n=0}^{\infty} (-1)^n x^n.$$

We can also differentiate and integrate the terms in a power series like we do with real functions. Consider the following theorem.

THEOREM: If $\sum_{n=0}^{\infty} a_n (x-a)^n$ has a radius of convergence $R > 0$, then

$f(x) = a_0 + a_1(x-a) + a_2(x-a)^2 + \ldots = \sum_{n=0}^{\infty} a_n (x-a)^n$ is differentiable and

continuous on $(a-R, a+R)$.

So, we know that

1. $f'(x) = \sum_{n=1}^{\infty} n a_n (x-a)^{n-1}$ and

2. $\int f(x)\,dx = \sum_{n=0}^{\infty} a_n \dfrac{(x-a)^{n+1}}{n+1} + C.$

A question does arise when dealing with power series. Is there a way we can write a function $f(x)$ as a power series? The answer is yes! Such a series is known as the Taylor Series.

TAYLOR SERIES

Let $f(x) = a_0 + a_1(x-a) + a_2(x-a)^2 + a_3(x-a)^3 + \ldots$, where $|x-a| < R$.

Our goal is to find the coefficients (a_n's) of the Taylor Series.

If $x = a$, then $f(a) = a_0$.

Now $f'(x) = a_1 + 2a_2(x-a) + 3a_3(x-a)^2 + \ldots$ So, $f'(a) = a_1$.

If $f''(x) = 2a_2 + 6a_3(x - a) + \ldots$ then $f''(a) = 2a_2$ or $a_2 = f''(a) / 2$.

Then $f'''(x) = 6a_3 + 24a_4(x - a) + \ldots$ So, $f'''(a) = 6a_3 = 3!a_3$ or $a_3 = f'''(a) / 3!$

If we continue this pattern with successive derivatives, then the nth coefficient is $a_n = f^{(n)}(a) / n!$ This result gives us the following theorem.

THEOREM: If $f(x) = \displaystyle\sum_{n=0}^{\infty} a_n (x - a)^n$ for $|x - a| < R$, then $a_n = \dfrac{f^{(n)}(a)}{n!}$.

This is known as the Taylor Series of the function $f(x)$.

If $a = 0$, then Taylor Series becomes $\displaystyle\sum_{n=0}^{\infty} \dfrac{f^{(n)}(0)}{n!} x^n$, which is called the MacLaurin Series.

Even though the Taylor Series is a good estimation of the function $f(x)$, the series itself does present some error into the equation. The next idea helps us tackle the remainder or error of the Taylor Series.

REMAINDER AND TAYLOR'S INEQUALITY

Suppose we let $T_n(x)$ be the Taylor Series of nth degree and let the remainder $R_n(x) = f(x) - T_n(x)$.

Then $f(x)$ is the sum of the Taylor Series if $f(x) = \displaystyle\lim_{n \to \infty} T_n(x)$ on $(a - R, a + R)$ and $\displaystyle\lim_{n \to \infty} R_n = 0$.

The next concept covers how we can estimate the error of the Taylor Series with Taylor's Inequality.

TAYLOR'S INEQUALITY: If $|f^{(n+1)}(x)| \leq N$ for $|x - a| \leq r$, then the remainder $R_n(x)$ of the Taylor Series satisfies the inequality

$$|R_n(x)| \leq \dfrac{N}{(n+1)!} |x - a|^{n+1}, \text{ whenever } |x - a| \leq r.$$

We can see why this inequality is true. Let us consider the case for $n = 1$.

If $f'(x) \leq N$, for $a - r \leq x \leq a + r$, then we have

$$\int_a^x f''(t)dt \leq \int_a^x Ndt \rightarrow f'(x) - f'(a) \leq N(x-a) \text{ or}$$

$$f'(x) \leq f'(a) + N(x-a).$$

We integrate again to get

$$\int_a^x f'(t)dt \leq \int_a^x [f'(a) + N(x-a)]dt, \text{ which implies that}$$

$$f(x) - f(a) \leq f'(a)(x-a) + \frac{N(x-a)^2}{2} \text{ or}$$

$$f(x) - f(a) - f'(a)(x-a) \leq \frac{N(x-a)^2}{2}.$$

On the left side of the inequality, we see that $T_1(x) = f(a) + f'(a)(x-a)$ and that $R_1(x) = f(x) - T_1(x)$. Thus, $R_1(x) \leq \dfrac{N(x-a)^2}{2}$.

Similarly, if we let $f''(x) \geq -N$, then $R_1(x) \geq \dfrac{-N(x-a)^2}{2}$.

It follows that $|R_1(x)| \leq \dfrac{N(x-a)^2}{2}$.

COMPUTATIONAL EXERCISES

1. Determine if each sequences converges or diverges. If the sequence converges, find the limit.

   a) $a_n = \dfrac{n+2}{2n-1}$

   b) $a_n = \dfrac{(n+1)!}{n!}$

   c) $a_n = \dfrac{(-1)^n \ln(3n)}{\ln(4n)}$

d) $a_n = \dfrac{\sin(n^2)}{n^3}$

2. Determine if the series converges or diverges. If the series converges, find the limit.

a) $\displaystyle\sum_{n=1}^{\infty} 6\left(\dfrac{2}{5}\right)^{n-1}$

b) $\displaystyle\sum_{n=1}^{\infty} \dfrac{3n-2}{4n+1}$

c) $\displaystyle\sum_{n=1}^{\infty} \dfrac{1}{n^2+9n+8}$

d) $\displaystyle\sum_{n=1}^{\infty} \left(\dfrac{7}{4^n} + \dfrac{3}{n(n+2)}\right)$

3. Use the integral test to determine if the series converges or diverges.

a) $\displaystyle\sum_{n=1}^{\infty} \dfrac{1}{n^3}$

b) $\displaystyle\sum_{n=1}^{\infty} \dfrac{1}{2n+4}$

c) $\displaystyle\sum_{n=2}^{\infty} \dfrac{1}{n\ln(n)}$

d) $\displaystyle\sum_{n=1}^{\infty} \dfrac{2}{n^2+1}$

4. Use the comparison test to determine if the series is convergent or divergent.

a) $\displaystyle\sum_{n=1}^{\infty} \dfrac{n+6}{n^2}$

b) $\displaystyle\sum_{n=1}^{\infty} \dfrac{1}{n^5+3n+7}$

c) $\displaystyle\sum_{n=0}^{\infty} \frac{3^n + 1}{8 + 4^n}$

5. Determine if the alternating series $\displaystyle\sum_{n=1}^{\infty} (-1)^n \frac{n}{n^2 + 1}$ is convergent.

6. Estimate the sum of the series $\displaystyle\sum_{n=1}^{\infty} \frac{(-1)^n}{n^2}$ if the error = 0.01.

7. Determine if the series is absolutely convergent, conditionally convergent, or divergent.

   a) $\displaystyle\sum_{n=1}^{\infty} \frac{n}{2^n}$

   b) $\displaystyle\sum_{n=1}^{\infty} \frac{(-1)^n}{n+1}$

   c) $\displaystyle\sum_{n=0}^{\infty} \frac{(3n)!}{3^n}$

   d) $\displaystyle\sum_{n=0}^{\infty} \frac{(-1)^n 4^n}{5^n (n+2)}$

8. Find the radius and interval of convergence for each series.

   a) $\displaystyle\sum_{n=1}^{\infty} \frac{x^n}{n^2}$

   b) $\displaystyle\sum_{n=1}^{\infty} \frac{x^n}{2^n n}$

   c) $\displaystyle\sum_{n=1}^{\infty} \frac{(x-1)^n}{3^n n!}$

9. Find a power series representation for each function.

   a) $f(x) = \dfrac{1}{1 + x^3}$

278

b) $f(x) = \dfrac{1}{x+4}$

c) $f(x) = \dfrac{x^2}{x+2}$

10. Use differentiation or integration to find a power series representation for each function.

a) $f(x) = \dfrac{1}{(1+x)^2}$

b) $f(x) = \ln(2-x)$

11. Find the Taylor series for $f(x)$ centered at $x = a$.

a) $f(x) = x^2 + 3x + 4$, $a = 2$

b) $f(x) = \ln x$, $a = 1$

c) $f(x) = \sin x$, $a = \pi / 4$

12. Find the MacLaurin series for $f(x) = e^x$. What is the interval and radius of convergence?

13. Use the MacLaurin series from exercise 12 to evaluate

$$\lim_{x \to 0} \dfrac{e^x - x - 1}{x^2}$$

14. Evaluate each integral as an infinite series.

a) $\displaystyle\int \dfrac{\cos x}{x} dx$

b) $\displaystyle\int \dfrac{e^x - 1}{x} dx$

15. If $f(x) = \sqrt{x}$ on the interval $|x - 2| \leq 2$, then use Taylor's Inequality to find $|R_2(x)|$.

CONCEPTUAL EXERCISES

1. True or False: If $\displaystyle\lim_{n \to \infty} |a_n| = 0$ then $\displaystyle\lim_{n \to \infty} a_n = 0$.

2. True or False: A sequence that is either increasing or decreasing is monotonic.

3. True or False: Every bounded sequence is convergent.

4. True or False: If $\sum a_n$ and $\sum b_n$ are convergent, then $\sum (a_n + b_n)$ is also convergent.

5. True or False: If $a_n \geq b_n$ and a_n is divergent, then b_n is also divergent.

6. True or False: If $f(x) > 0$ and decreasing on $[1, \infty)$, then $\sum f(n)$ is convergent.

7. True or False: If $\displaystyle\lim_{n \to \infty} a_n \neq 0,$ then $\sum a_n$ diverges.

8. True or False: Every Taylor series also has a MacLaurin series and vice versa.

9. True or False: The sum of a convergent series and a divergent series is also divergent.

10. True or False: The MacLaurin series for e^x is convergent for all x.

11. Prove that $\displaystyle\lim_{n \to \infty} \frac{n+1}{n} = 1.$

12. Show that $a_n = \dfrac{2n}{n^2 + 4}$ is monotonic for all $n \geq 1$.

13. Find a recursive formula for the sequence $\sqrt{\pi}, \sqrt{\pi\sqrt{\pi}}, \sqrt{\pi\sqrt{\pi\sqrt{\pi}}}, \ldots$
    What is the limit of this sequence?

14. If $\lim\limits_{n \to \infty} a_n = A$ and $\lim\limits_{n \to \infty} b_n = B$,

    then show that $\lim\limits_{n \to \infty}(a_n + b_n) = A + B$.

15. Use the integral test to prove that if $\sum\limits_{n=1}^{\infty} \dfrac{1}{n^p}$ is convergent, then $p > 1$.

16. Prove or disprove. If both $\sum a_n$ & $\sum b_n$ are divergent, then
    $\sum(a_n + b_n)$ is also divergent.

17. Find a power series for the function $f(x) = \dfrac{1}{b - ax}$. What is the
    interval and radius of convergence?

18. Show that $2\cosh x \geq x^2 + 2$ for all x. Hint: Recall that
    $\cosh x = \dfrac{e^x + e^{-x}}{2}$.

19. Prove or disprove. $\sum\limits_{n=1}^{\infty} \dfrac{n^n}{n!}$ is a convergent series.

20. Prove Taylor's Inequality for $n = 3$.